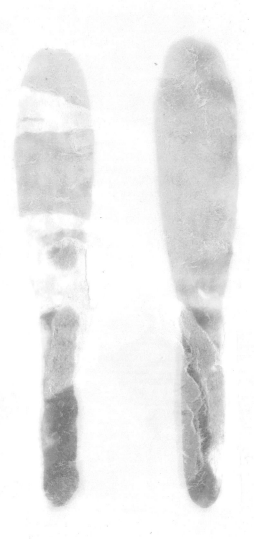

The
Viruses

CATALOGUE, CHARACTERIZATION,
AND CLASSIFICATION

THE VIRUSES

Series Editors
HEINZ FRAENKEL-CONRAT, *University of California*
Berkeley, California

ROBERT R. WAGNER, *University of Virginia School of Medicine*
Charlottesville, Virginia

THE VIRUSES: Catalogue, Characterization, and Classification
Heinz Fraenkel-Conrat

THE ADENOVIRUSES
Edited by Harold S. Ginsberg

THE HERPESVIRUSES,
Volumes 1–3 • Edited by Bernard Roizman
Volume 4 • Edited by Bernard Roizman and Carlos Lopez

THE PARVOVIRUSES
Edited by Kenneth I. Berns

THE PLANT VIRUSES
Volume 1 • Edited by R. I. B. Francki

THE REOVIRIDAE
Edited by Wolfgang K. Joklik

The Viruses

CATALOGUE, CHARACTERIZATION, AND CLASSIFICATION

HEINZ FRAENKEL-CONRAT

Department of Molecular Biology and Virus Laboratories
University of California
Berkeley, California

PLENUM PRESS • NEW YORK AND LONDON

Library of Congress Cataloging in Publication Data

Fraenkel-Conrat, Heinz, 1910–
   The viruses: catalogue, characterization, and classification.

   Includes bibliographies and index.
   1. Viruses. I. Title. [DNLM: 1. Viruses. QW 160 F799va]
QR360.F715   1985   576′.64   84-26649
ISBN 0-306-41766-9

© 1985 Plenum Press, New York
A Division of Plenum Publishing Corporation
233 Spring Street, New York, N.Y. 10013

Printed in the United States of America

Preface

During the past two decades, virus taxonomy has advanced to the point where most viruses can be classified as belonging to families, genera, or groups of related viruses. Virus classification is primarily based on chemical and physical similarities, such as the size and shape of the virion, the nature of the genomic nucleic acid, the number and function of component proteins, the presence of lipids and of additional structural features, such as envelopes, and serological interrelationships. The families, genera, or groups of viruses that have been defined on the basis of such criteria by the International Committee on Taxonomy of Viruses (ICTV) will be described in some detail in this catalogue and illustrated by electron micrographs. In my present attempt to list most if not all well established and studied viruses in alphabetical order, I have largely confined myself to identifying them only in such taxonomic terms, generally without quoting specific data reported for individual viruses. If the latter data do not at times agree closely with those given for the taxon or group, it is difficult to decide to what extent this is attributable to misclassification due to insufficient data and errors in the analytical procedures and descriptions, or to what extent this is an expression of Nature's freedom of choice and abhorrence of restrictive classifications.

The classification of the animal viruses, including protozoa, that is dealt with in Section I of this book uses families (presently 19, named with the ending -viridae), subfamilies for three of these (ending in -virinae), and genera; species have in most cases not yet been officially identified. The taxonomic significance of terms such as *strains, mutants,* and *(sero)types* is not clearly established. Only those of the latter that have been studied in detail are listed.

Most plant viruses are only classified in terms of groups, with many viruses remaining uncertain in terms of classification or identification as separate viruses, rather than strains. The plant viruses, including those of protophyta, are dealt with in Section II. However, it must be noted that the separate listing of plant and animal viruses is frequently arbitrary, since many plant viruses replicate in their insect (or other) vector. Those that are pathogenic in their plant host will be listed as plant viruses.

Family names have been coined and officially adopted for many of the bacterial and blue-green algal viruses, but no further classification has been generally accepted. Most of these viruses are identified only by letters or numbers, or by the host from which they were isolated, the same "names," i.e., identifying letters and/or numbers, frequently recurring in unrelated phages. Also many phage isolates with different "names" may actually be identical. Thus, it is questionable whether an attempt to list the phages in alphabetical order is worthwhile. Section III covers mostly bacteriophages that have been studied and used repeatedly and in more than one laboratory. To the extent possible, these are identified in terms of their family status, recognizing that that classification scheme is less advanced than that of the animal viruses.

The (deoxy) nucleotide and amino acid sequences of very many virus components have been and continue to be currently established. Thus the references to known sequences in this catalogue are incomplete. Such sequences are now known, at least in part, for most of the important viruses.

The number of citations that would be required to support the description of each virus would at least triple the size of this book. The use of a single "key" reference for each virus appears arbitrary and not always helpful. I therefore reference mainly the virus families and groups and, whenever possible, quote review chapters and occasionally recent papers to assist the reader in further searches. Also helpful are *Classification and Nomenclature of Viruses*, the Fourth Report of the International Committee on Taxonomy of Viruses (R. E. F. Matthews, Intervirology **17**:1–3, 1982), *Virology Abstracts* (Cambridge Scientific Abstracts), the cumulative indexes of *Virology*, the *Journal of Virology*, and the *Journal of General Virology*, the recent textbook *Virology* (H. Fraenkel-Conrat and P. C. Kimball, Prentice-Hall, Englewood Cliffs, New Jersey, 1982), and the monograph *Plant Virology*, Second Edition (R. E. F. Matthews, Academic Press, New York, 1981). Most plant viruses are described in detail in the CMI/AAB *Description of Plant Viruses* series of pamphlets, edited by A. F. Murant and B. D. Harrison. Obviously, other volumes of this series (*The Viruses*), as they appear, will represent the most up-to-date source of information on each virus family or group. Currently available or in preparation are books on the herpesviruses (four volumes), the Reoviridae, the parvoviruses, the adenoviruses, etc.

Most virus families or groups of characteristic shapes are illustrated by electron micrographs. I am greatly indebted to Dr. R. C. Williams for Figures 1–12, 14–17, 19–25, 32, 34–36, and 42. Figure 13 was kindly supplied by Dr. D. W. Verwoerd, Figure 18, by Dr. E. M. J. Jaspars, Figure 40 by Dr. M. Salas, and the others by Dr. H.-W. Ackermann.

Heinz Fraenkel-Conrat

Contents

Animal Viruses, Including Protozoal Viruses

Virus	Group or subgroup	Genus	Subfamily or family
Abadina (transmitted by *Culicoides*)	Palyam	orbivirus	Reoviridae
Abelson's murine leukemia[1]: *see* oncoviruses			
Abras	probably Patois	bunyavirus	Bunyaviridae
Abraxas grossularia	type 8	cypovirus	Reoviridae
Abu Hammad	Dera Ghazi Khan	nairovirus	Bunyaviridae
Abu Mina	Dera Ghazi Khan	nairovirus	Bunyaviridae
Acado (mosquito transmitted)	Corriparta	orbivirus	Reoviridae
Acara	Capim	bunyavirus	Bunyaviridae
Acelaphine herpes h1: *see* malignant cattarrhal fever of wildebeest			
Acelaphine herpes h2: *see* hartebeest herpes virus			
Acheta		probably densovirus	Parvoviridae
Acrobasis zelleri (*Lepidoptera*)		B	Entomopoxvirinae
Actias selene	type 4	cypovirus	Reoviridae
Acute bee paralysis: *see* bee acute paralysis			
Acute hemorrhagic conjunctivitis (EV serotype 70; also 69, 71)		enterovirus	Picornaviridae
Acute infectious lymphocytosis (see EVU-16)[2]			
Acute laryngotracheobronchitis	type 2	parainfluenza	Paramyxoviridae
Adenoassociated[3]: synonym for dependovirus			Parvoviridae

Virus	Group or subgroup	Genus	Subfamily or family
ADENOVIRIDAE[4]: A very uniform family of icosahedral nonenveloped virions of about 80 nm diameter composed of 252 capsomers, buoyant density in CsCl 1.34 g/cm³. The 12 vertex capsomers (pentons) carry strain-characteristic glycoprotein fibers (10–30-nm long) with knobs at the ends; the rest is made up of hexons. The genome is linear double-stranded DNA of 20–25 × 10⁶ daltons in the mammalian and 28–30 × 10⁶ daltons in the avian viruses. At least ten proteins of 5–120 × 10³ daltons make up the virion. Each adenovirus has a narrow host range. Several are oncogenic in newborn nonhosts. Serological relations among the many different adenoviruses usually called "types" are very limited. Two genera have been defined (MASTADENOVIRUS and AVIADENOVIRUS; subgroups A–D or I–IV), and species names for those found in various animals have been proposed with h1–h34 for the (sero)types of the human virus species. Similar differentiation of serotypes are used for the animal adenoviruses (27 simian, ten bovine, eight avian, four porcine, two canine, and one ovine and opossum in 1979). Adenoviruses generally cause only light upper respiratory diseases (Figure 1).			
Adoxophyses orana[5]	nuclear polyhedrosis virus (A)		Baculoviridae
Aedes			Togaviridae
Aedes aegypti		probably densovirus	Parvoviridae
Aedes aegypti (Diptera)		C	Entomopoxvirinae
Aedes cantans		probably chloriridovirus	Iridoviridae
Aedes iridescent: see mosquito iridescent virus			
Aerocystis agent: see swim bladder inflammation agent of carp			
African green monkey cytomegalo (AGM-CMV)	cercopithecine	(h5) cytomegalovirus	Betaherpesvirinae
African green monkey EBV-like (AGM-EBV)	cercopithecine	(h14) lymphocrytovirus	Gammaherpesvirinae

African horse sickness (vector *Culicoides* spp.)	9 serotype	orbivirus	Reoviridae
African swine fever[6]	possibly genus of Iridoviridae (only five proteins)		
AG 80–24	probably Anopheles A	bunyavirus	Bunyaviridae
Agrophylla lutenta	type 10	cypovirus	Reoviridae
Agrotis segetum	type 9	cypovirus	Reoviridae
Aguacate		phlebovirus	Bunyaviridae
AIDS (acquired immune deficiency syndrome): *see* Human T-cell leukemia virus			
Aino	Simbu	bunyavirus	Bunyaviridae
Akabane	Simbu (RNA: 31 S, 26 S, 13 S)	bunyavirus	Bunyaviridae
AKR: *see* oncovirus			
AKv (mouse): *see* oncovirus			
Alajeula	probably Gamboa	bunyavirus	Bunyaviridae
Alastrim	identical to variola minor	orthopoxvirus	Chordopoxvirinae
Aleuquer		phlebovirus	Bunyaviridae
Aleutian disease of mink[7]	(nontypical proteins), (causes immune-complex disease)		possibly Parvoviridae

Virus	Group or subgroup	Genus	Subfamily or family
Alfuy (mosquito-borne)		flavivirus	Togaviridae
Allerton: *see* bovine ulcerative mammilitis			
Alpha: genus of Togaviridae[8] (type species: Sindbis virus)			
Alphaherpesvirinae[9]: subfamily of Herpesviridae; rapidly growing, highly cytolytic			
Amapari	Tacaribe complex		Arenaviridae
Amphibian cytoplasmic		ranavirus	Iridoviridae
Amsacta moorei (*Lepidoptera*)	type species	B	Entomopoxvirinae
ana 1 (*Anas domestica*, duck)		aviadenovirus	Adenoviridae
Ananindena	Guama	bunyavirus	Bunyaviridae
Anatid herpes, h1: *see* duck plaque herpesvirus			
Anhanga		phlebovirus	Bunyaviridae
Anhembi	Bunyamwera	bunyavirus	Bunyaviridae
Anomala cuprea (*Coleoptera*)		A	Entomopoxvirinae
Anopheles A	type species of subgroup	bunyavirus	Bunyaviridae
Anopheles B	type species of subgroup	bunyavirus	Bunyaviridae
ans 1 (*Anser domesticus*, goose)		aviadenovirus	Adenoviridae
Antheraea eucalypti	Nudaurelia β group		

Aotine h1, 2, 3: *see* herpes aotusviruses

Aotine h4: *see* owl monkey cytomegalovirus

Apanteles melanoscelus (wasp)	probably (double-stranded circular DNA of 2–25 × 10^6 daltons) (D)		Baculoviridae
Apeu	C	bunyavirus	Bunyaviridae
Aphid viruses[10] (*Rhopalosiphum padi*)	27 nm diameter (162 S), buoyant density CsCl 1.37 g/cm^3, ss RNA (31 S)		possibly Nodaviridae
Aphodius tasmaniae (*Coleoptera*)		A	Entomopoxvirinae

Aphtho[11]: genus of Picornaviridae (type species: foot and mouth disease virus, aphthovirus D)

Apoe		flavivirus	Togaviridae
Aporophylla lutulenta	type species of type 10	cypovirus	Reoviridae

Arbo: obsolete term for many *arthropod-borne* toga etc. viruses

ARENAVIRIDAE[12]: Enveloped pleomorphic though predominantly round virions of 100–200 nm diameter and 325–580 S, 1.2 g/cm^3 density in sucrose, consisting of a core containing ribosome-like particles (arena = sand) and a lipid bilayer envelope with surface projections. Two viral minus-stranded RNAs of about 1.1 and 2.7 × 10^6 daltons and smaller ribosomal etc. RNAs are present. The nucleocapsid protein is about 63 × 10^6 daltons; about two glycoproteins are present. The host range of each arenavirus is narrow. Members of the family show more or less close serological relationships. The type species is lymphocytic choriomeningitis virus (LMC), others the Lassa and Mozambique viruses and the American Tacaribe complex with many members. Several are pathogenic to man and young rodents, and LMC generally persistent. Transmission is vertical and horizontal, often venereal (Figure 2).

Virus	Group or subgroup	Genus	Subfamily or family
Argentina: Strain of vesicular stomatitis virus, closely related to cocal virus			
Arkansas bee[13]	(41 × 10³ dalton protein, yet probably nodavirus)		Nodaviridae
Arumowot		phlebovirus	Bunyaviridae
Asia-1		aphthovirus	Picornaviridae
Astro[14]	unclassified 28 nm isometric particles consisting of two proteins of 33 × 10³ daltons, and 2.8 × 10⁶ dalton RNA carrying 3' terminal poly(A)		
Ateline herpes 1: *see* spider monkey cytomegalovirus			
Ateline herpes 2, 3: *see* herpes ateles virus			
ATS-124: *see* oncovirus			
Aura (related to western equine encephalitis virus)		alphavirus	Togaviridae
Aus MK 6357 (transmitted by mosquitoes)		orbivirus	Reoviridae
Australian antigen: circulating hepatitis B virus protein aggregate			
Autographa californica[15] (wide host range pesticide)	nuclear polyhedrosis virus (A)		Baculoviridae
Avalon	Sakhalin	nairovirus	Bunyaviridae
Aviadenoviruses: genus of Adenoviridae (of fowl, goose, duck, pheasant)			
Avian adenoviruses: *see* aviadenoviruses			

Avian C-type: *see* oncovirus

Avian dependoviruses: *see* adenoassociated viruses

Avian encephalomyelitis — probably cardiovirus — Picornaviridae

Avian erythroblastosis: *see* oncovirus

Avian herpes: *see* Marek's disease

Avian herpes, h1 herpesviruses: Anatid (duck), ciconid (stork), columbid (pigeon), gruid (crane), meleagrid (turkey), perdicid (quail), phalacrocoracid (cormorant), psittacid (parrot), strigid (owl); all as far as known Alphaherpesvirinae.

Avian infectious bronchitis (IBV)[16] — type species of coronavirus — Coronaviridae

Avian influenza A — orthomyxovirus — Orthomyxoviridae

Avian leukemia: *see* oncovirus

Avian leukosis: *see* oncovirus

Avian myeloblastosis (AMV) and myeloblastosis-associated: *see* oncovirus

Avian myelocytomatosis (MC 29): *see* oncovirus

Avian osteopetrosis: *see* oncovirus

Avian reticuloendotheliosis: *see* oncovirus

Avian sarcoma[17]: *see* oncovirus

Avipox[18]: subgroup of Chordopoxvirinae (type species: fowlpox virus)

B (monkey herpes) (h1) — simplexvirus — Alphaherpesvirinae

B 19 (pathogenic in children) — parvovirus — Parvoviridae

Virus	Group or subgroup	Genus	Subfamily or family
B 77 (chicken): *see* oncovirus			
B 1327	blue tongue	orbivirus	Reoviridae
Babahoyo	Patois	bunyavirus	Bunyaviridae
Baboon herpes (h12)		lymphocrytovirus	Gammaherpesvirinae
Baboon sarcoma: *see* oncovirus			

BACULOVIRIDAE[19]: The virions are usually rod-shaped nucleocapsids (*baculus* = stick), with lipid bilayer envelopes, frequently in bundles occluded in "crystalline" protein bodies. The nucleocapsids are about 50 × 300 nm and have a density in CsCl of 1.47 g/cm³, compared to 1.21 for the enveloped virion. The genomes are circular double-stranded DNAs of 60–110 × 10⁶ daltons, and there are 10–25 proteins, including the single virus-coded matrix, termed polyhedrin, for the NUCLEAR POLYHEDROSIS SUBGROUP (A) and granulin for the GRANULOSIS SUBGROUP (B). The viruses occur in insects, spiders, and crustaceans. The nucleocapsids of the nuclear polyhedrosis group are frequently multiply enveloped, and always have many virions in each occlusion body; the granulosis virions occur singly. Proposed are two subgroups of nuclear nonoccluded enveloped rod-shaped virions (C) and (D), the latter characterized by polydisperse superhelical DNA in various length particles. The proteins of the occlusion bodies of all genera are serologically related. Transmission is horizontal and vertical (through the eggs) (Figure 3).

Virus	Group or subgroup	Genus	Subfamily or family
Baculo X	nonoccluded, singly enveloped persistent (C)		Baculoviridae
Bagaza	(mosquito-borne)	flavivirus	Togaviridae
Bahia Grande (TB4 1054)	(invertebrate hosts)		probably Rhabdoviridae
Bahig	Tete	bunyavirus	Bunyaviridae

	type species of possible subgroup		
Bakau		bunyavirus	Bunyaviridae
Baku (tick-transmitted)	Kemorovo	orbivirus	Reoviridae
Balb/2 (murine): *see* oncovirus			
Balb-10-1: *see* oncoviruses			
Bambari (mosquito-transmitted)	Corriparta	orbivirus	Reoviridae
Banded Krait (herpes)			Herpesviridae
Bandia	Qalyub	nairovirus	Bunyaviridae
Banzi	(mosquito-borne)	flavivirus	Togaviridae
Barathra brassica[5]	nuclear polyhedrosisvirus (A)		Baculoviridae
Barmah Forest		alphavirus	Togaviridae
Barur	(vertebrate and invertebrate hosts)		probably Rhabdoviridae
Batai	Bunyamvera	bunyavirus	Bunyaviridae
Batama	Tete	bunyavirus	Bunyaviridae
Bat salivary: *see* salivary bat virus			
Batu Cave		flavivirus	Togaviridae
Bauline (tick-transmitted)	Kemorovo	orbivirus	Reoviridae
Be An 157575 (vertebrate hosts)			probably Rhabdoviridae

Virus	Group or subgroup	Genus	Subfamily or family
Be An 293022			probably Arenaviridae
Be Ar 35646, 41067, 54342 (transmitted by phlebotomines)	Changuinola	orbivirus	Reoviridae
Be Ar 185559 (invertebrate hosts)			probably Rhaboviridae
Bebaru		alphavirus	Togaviridae
Bee acute paralysis[20]	unclassified (very similar to sacbrood virus, possibly rhinovirus, but not serologically related, stable < pH 4, 28 nm diameter, isometric particles 160 S, proteins of 24 and 32 × 10^3 daltons)		
Bee chronic paralysis[20]	unclassified (particles of 20 × 30, 40, 50, 60 nm, 82–126 S, 66 × 10^3 dalton protein, often carries isometric satellite virus of 17 nm diameter)		
Bee filamentous[20]	unclassified (ellipsoid virions, 150 × 450 nm, density in CsCl 1.28 g/cm³, contains 40 × 3000 nm nucleocapsid, double-stranded DNA of 12 × 10^6 daltons and about 12 proteins of 13–70 × 10^3 daltons)		
Bee slow paralysis[20]	unclassified (30 nm isometric particles, 178 S, proteins of 46, 29, and 27 × 10^3 daltons, plus-strand RNA)		
Bee X and Y[20]	(very similar, but serologically distantly related, 35-nm-diameter isometric particles, 187 S, 55 × 10^3 dalton protein, probably Nudaurelia β group)		

Name	Notes / serogroup	Genus	Family
Belmont[21]	(isolated only in Australia, somewhat larger virion and larger three RNAs, four proteins)		probably Bunyaviridae
Benevides	Capim	bunyavirus	Bunyaviridae
Bentica	Capim	bunyavirus	Bunyaviridae
Berne[22]	Equine unclassified 130-nm-diameter particle with 20-nm spikes, partly bacilliform (buoyant density in sucrose 1.16 g/cm³, contains RNA. The structure of the core differentiates it from coronaviridae, not pathogenic)		
Bertioga	Guama	bunyavirus	Bunyaviridae
Betaherpesvirinae[23]: subfamily of Herpesviridae (slow growing, cytomegalic)			
BFN 3187 (Grey Lodge) (invertebrate hosts)			probably Rhabdoviridae
Bhanja			probably Bunyaviridae
Bijou bridge		alphavirus	Togaviridae
Bimiti	Guama	bunyavirus	Bunyaviridae
Birao	Bunyamwera	bunyavirus	Bunyaviridae
Bird pox: see avipox			

BIRNA VIRUS GROUP[24]: Not yet classified family; icosahedral 60-nm-diameter particles (at times resembling reoviridae), 435 S, density in CsCl 1.32 g/cm³, contain two linear double-stranded RNAs of 2.5 and 2.3 × 10⁶ dalton per particle, and four proteins of 105 to 29 × 10³ daltons. Type species: infectious pancreatic necrosis of trout; viruses of fishes, molluscs, possibly *Drosophila* X, infectious bursal disease of chickens. Wide host range, three serotypes. The RNA may be circularized by a protein.

Virus	Group or subgroup	Genus	Subfamily or family
Biston betularia	type 6	cypovirus	Reoviridae
Bittner: *see* mouse mammary tumor virus			
BK (human)[25]	Related to SV40 (3.45 × 10⁶ dalton circular dsDNA, 4962 × 2 nucleotides, sequenced).	polyoma	Papovaviridae
Black beetle[26]		nodavirus	Nodaviridae
Black stork herpes (h1)	ciconiid		Herpesviridae
Blue comb disease (enteritis of turkeys)		coronavirus	Coronaviridae
Blue tongue: Type species of orbivirus—68-nm-diameter particles, 550 S, two shells, seven proteins of 155 to 32 × 10³ daltons, ten double-stranded RNAs of 2.7 to 0.3 × 10⁶ daltons. Sensitive to lipid solvents and pH 3, many serotypes and serological subgroups, wide host range, mostly arthropods and mammals, various transmitting insects.			Reoviridae
Bobia	Olifantsvlei	bunyavirus	Bunyaviridae
Bobwhite quail herpes	perdicid	herpesvirus	Herpesviridae
Bombyx mori		densovirus	Parvoviridae
Bombyx mori	nuclear polyhedrosis (A)		Baculoviridae
Bombyx mori	type 1	cypovirus	Reoviridae

Boolarra			possibly Nodaviridae
Boraceia	Anopheles B	bunyavirus	Bunyaviridae
Border disease		pestivirus	Togaviridae
Borna disease	unclassified		Togaviridae
Borrelina bombycis	nuclear polyhedrosis (A)		Baculoviridae
bos 1–9 (Bos taurus, cattle)		mastadenovirus	Adenoviridae
Botambi	Olifantsvlei	bunyavirus	Bunyaviridae
Bouboni (mosquito-transmitted)		flavivirus	Togaviridae
Bovid: see bovine			
Bovine adeno		adenovirus	Adenoviridae
Bovine dependo	(adeno-associated virus)	dependovirus	Parvoviridae
Bovine diarrhea		probably enterovirus	Picornaviridae
Bovine entero 1–7	VG-5-27	enterovirus	Picornaviridae
Bovine ephemeral (or epizootic) fever[27]			probably Rhabdoviridae

Bovine herpes h1[28]: see infectious bovine rhinotracheitis

Bovine herpes h2[28]: see bovine ulcerative mammillitis, pseudolumpy skin disease, Allerton virus

Bovine herpes h3[28]: see bovine "orphan" herpes virus

Virus	Group or subgroup	Genus	Subfamily or family
Bovine leukosis: *see* oncovirus			
Bovine lumpy disease		parapoxvirus	Chordopoxvirinae
Bovine mammilitis[29]: *see* bovine ulcerative mammilitis			
Bovine orphan (herpes, h2)		simplexvirus	Alphaherpesvirinae
Bovine papilloma	type 1 (sequenced)		Papovaviridae
Bovine parainfluenza	type 3	parainfluenza	Paramyxoviridae
Bovine parvo		parvovirus	Parvoviridae
Bovine pustular stomatitis		parapoxvirus	Chordopoxvirinae
Bovine respiratory syncytial		pneumovirus	Paramyxoviridae
Bovine rhino 1, 2		rhinovirus	Picornaviridae
Bovine rhinotracheitis: *see* infectious bovine rhinotracheitis			
Bovine syncytial			Spumavirinae
Bovine ulcerative mammilitis (Allerton virus), (h12)		simplexvirus	Alphaherpesvirinae
Bovine viral diarrhea[30]	type species	pestivirus	Togaviridae
Bratislava (77): *see* avian sarcoma viruses			
Brazil: strain of vesicular stomatitis virus			Rhabdoviridae
Brazilian myxoma		leporipoxvirus	Chordopoxvirinae
Bruconha	probably C	bunyavirus	Bunyaviridae

Bryan strain of Rous sarcoma virus: *see* oncovirus

BT 104	Changuinola	orbivirus	Reoviridae
BT 2164	Changuinola	orbivirus	Reoviridae

B-type particles: *see* Oncovirus

Buenaventura		phlebovirus	Bunyaviridae
Buffalopox		orthopox	Chordopoxvirinae
Bujaro		phlebovirus	Bunyaviridae
Bukalasa bat		flavivirus	Togaviridae
Bunyamwera	type species	bunyavirus	Bunyaviridae

Bunyamwera supergroup: synonym for genus Bunyavirus

BUNYAVIRIDAE[31]: Large family of viruses. Oval to spherical virions of about 95 nm diameter (400 S, density in CsCl 1.2 g/cm³) containing three major envelope glycoproteins, one minor large nucleocapsid protein forming long helical nucleocapsids (2.0–2.5 nm diameter) with the three minus strand RNAs (3–5, 1–2, and 0.3–0.8 × 10⁶ daltons). Reassortment between the RNAs of different bunyaviruses or strains that carry the information for the proteins of equivalent size has been achieved. The viruses have lipid-rich envelopes. Their host ranges are wide among vertebrates and anthropods; transmission is usually by mosquitoes, ticks, etc. Four genera have been identified (BUNYAVIRUS, PHLEBOVIRUS, NAIROVIRUS, and UUKUVIRUS), as well as many groups for each of the first three genera. The phleboviruses have the largest small RNA, the bunyaviruses the smallest (0.8 vs. 0.4 × 10⁶ daltons); their large RNA exceeds that of the others (4.6 × 10⁶ daltons).

Bunyavirus: genus of Bunyaviridae (type species: bunyamweravirus)

Bushbush		Bunyaviridae
Capim	bunyavirus	

Virus	Group or subgroup	Genus	Subfamily or family
Bussuquari (mosquito-transmitted)		flavivirus	Togaviridae
Button willow	Simbu	bunyavirus	Bunyaviridae
Butus occitanns (scorpion)			probably Reoviridae
Bwamba	type species of subgroup	bunyavirus	Bunyaviridae
C: subgroup		bunyavirus	Bunyaviridae
C-type: *see* oncovirus			
C 57L, C 58 (murine leukemia virus): *see* oncovirus			
Cabassou		alphavirus	Togaviridae
Cacao		phlebovirus	Bunyaviridae
Cache valley	Bunyamwera	bunyavirus	Bunyaviridae
Caimito		phlebovirus	Bunyaviridae
Calf rota[32]		rotavirus	Reoviridae

CALICIVIRIDAE[33]: This virus family shows some resemblance to enlarged picornaviridae. The isometric particles have a diameter of 38 nm (183 S), density 1.37 g/cm^3 in CsCl), and contain 18% plus-strand RNA of 2.8×10^6 daltons with poly (A) at the 3' end and a $10-15 \times 10^3$ dalton protein bound to the 5' end. However, they are composed of 180 molecules of a single protein of about 67×10^3. Their particle weight is thus 15×10^6, compared to 8×10^6 for the picornaviruses. The type species of this small family is the vesicular exanthema of swine virus with many serotypes, and others are the San Miguel sea lion virus, a feline calici virus, and others. These show serological interrelationships.

California diarrhae		rotavirus	Reoviridae

California encephalitis[34]	type species of subgroup	bunyavirus	Bunyaviridae
California: (subgroup)		bunyavirus	Bunyaviridae
California myxoma		leporipox	Chorodopoxvirinae
Callinectes sapidus (blue crab)			possibly Baculoviridae
Callitrichine herpes h1: *see* herpes virus sanguinus			
Callitrichine herpes h2: *see* SSG, marmoset cytomegalovirus			
Calovo	Bunyamwera	bunyavirus	Bunyaviridae
Camel pox		orthopox	Chordopoxvirinae
Campoletis sonorensis (wasp)	probably D		Baculoviridae
Camptochironomus tentans (*Diptera*)		B	Entomopoxvirinae
can-1	(*Canis familiaris*, dog)	mastadenovirus	Adenoviridae
Cananeia	Guama	bunyavirus	Bunyaviridae
Canary pox		avipox	Chordopoxvirinae
Candiru		phlebovirus	Bunyaviridae
Canid: *see* canine			
Canine corona		coronavirus	Coronaviridae
Canine depende[35] (related to others)		dependovirus	Parvoviridae
Canine distemper		morbillivirus	Paramyxoviridae

Virus	Group or subgroup	Genus	Subfamily or family
Canine hepatitis			possibly Alphaherpes-virinae
Canine herpes: *see* dog herpesvirus			
Canine papilloma		papillomavirus	Papovaviridae
Canine parvo: *see* canine dependovirus			
cap-1	(*capra hercus*, goat)	mastadenovirus	Adenoviridae
Cape Wrath (tick-transmitted)	Kemorovo	orbivirus	Rhabdoviridae
Caprine arthritis-encephalitis[36]	(visna-like serology but no RNA homology)		probably Lentivirinae
Caprine herpes: *see* domestic goat herpes virus, sheep herpes			
Caprine herpes h1: *see* sheep herpes virus			
Caprine herpes h2: *see* domestic goat herpes virus			
Capripox[37]: genus of Chordopoxvirinae (type species: sheeppox)			
Capuchin herpes (AL-5, AP18) (h1, h2)		simplexvirus h1, h2	Alphaherpesvirinae
Caraparu	C	bunyavirus	Bunyaviridae
Carcinoma MH2: *see* oncovirus			

Carcinus maenas (European crab)			possibly Baculoviridae
Cardio[38]: genus of Picornaviridae (type species: encephalomyocarditis virus)			
Carey Island		flavivirus	Togaviridae
Carnivorepox		orthopox (related to cowpox)	Chordopoxvirinae
Carppox			Herpesviridae
Carr–Zilber strain of RSV, and associated: *see* oncovirus			
Cas-Br-M-(wild mouse): *see* oncovirus			
Cas E no 1-X: *see* oncovirus			
Cas SFFV (mouse): *see* oncovirus			
Cat cytomegalo (h2)		cytomegalovirus	Bethaherpesvirinae
Cat herpes (infectious rhinotracheitis)			Alphaherpesvirinae
Cat scratch disease		unclassified	
Catu	Guama	bunyavirus	Bunyaviridae
Caviid herpes 1: *see* guinea pig herpes virus, Hsuing–Kaplow			
Caviid herpes h2: *see* guinea pig cytomegalovirus			
Cba Ar (426)	Bunyamwera	bunyavirus	Bunyaviridae
CELO		aviadenovirus	Adenoviridae

Virus	Group or subgroup	Genus	Subfamily or family
Ceratitis	similar to *Drosophila* virus F		
Cercopithecine herpes h1: *see* herpes B, simian herpes B; h2: *see* SA8; h3: *see* SA6; h4: *see* SA 15; h5: *see* African green monkey cytomegalo; h6: *see* Liverpool vervet monkey; h7: *see* delta herpes, Patas monkey; h8: *see* Rhesus monkey cytomegalo; h9: *see* simian varicella, Medical Lake macaque; h10: *see* rhesus leukocyte-associated (strain I); h11: *see* rhesus leukocyte-associated (strain II); h12: *see* baboon herpes, herpes papio; h13: *see* herpes cyclopsis; h14: *see* African green monkey EBV-like.			
CFS (cell-fusing agent)	(though serologically unrelated to others)	flavivirus	Togaviridae
C: subgroup		bunyavirus	Bunyaviridae
CH 9935		orbivirus	Reoviridae
Chaco (vertebrate hosts)			probably Rhabdoviridae
Chagres		phlebovirus	Bunyaviridae
Chamois contagious ecthyma		parapox	Chordopoxivirinae
Chandipura (human)		vesiculovirus	Rhabdoviridae
Changuinola (transmitted by *phlebotomines*)	type species of subgroup (7 serotypes)	orbivirus	Reoviridae
Channel catfish[39] (h1)			Alphaherpesvirinae
Chelonid herpes h1: *see* grey patch disease agent of green turtle			
Chelonid herpes h2: *see* Pacific pond turtle			

Chelonid herpes h3: *see* painted turtle herpes			
Chenuda (tick-transmitted)	Kemorovo	orbivirus	Reoviridae
Chickenpox: *see* varicella-zoster			
Chikungunya		alphavirus	Togaviridae
Chilibre		phlebovirus	Bunyaviridae
Chimpanzee herpes (h1)		lymphocrytovirus	Gammaherpesvirinae
Chironomus attenuatus (*Diptera*)	type species	C	Entomopoxvirinae
Chironomus luridus (*Diptera*)		C	Entomopoxvirinae
Chironomus plumosus (*Diptera*)		C	Entomopoxvirinae
Chlorirido[40]: genus of Iridoviridae	(larger than iridoviruses)		
Chordopoxvirinae[41]: subfamily of Poxviridae (of vertebrates)	Subfamily of Poxviridae		
Choristoneura biennis (*Lepidoptera*)		B	Entomopoxvirinae
Choristoneura conflicta (*Lepidoptera*)		B	Entomopoxvirinae
Choristoneura diversuma (*Lepidoptera*)		B	Entomopoxvirinae
Chorizagrotis auxiliaris (*Lepidoptera*)		B	Entomopoxvirinae

Virus	Group or subgroup	Genus	Subfamily or family
Chrysodeixis eriosoma		cotiavirus	Chordopoxvirinae
Chum salmon	similarities to (11 ds RNAs) and differences from	rotavirus	Reoviridae
Ciconiid herpes h1: *see* black stork herpes virus			
Clo Mor		nairovirus	Bunyaviridae
CMII: *see* avian oncovirus (member of the MC 29 group)			
Co Ar 1071, Co Ar 3624, Co Ar 3627	Anopheles A	bunyavirus	Bunyaviridae
Co Ar 2837	Changuinola	orbivirus	Reoviridae
Cocal: strain of vesicular stomatitis virus			
Coital exanthema (horse) (h3)			Alphaherpesvirinae
Col An 57389	Anopheles A	bunyavirus	Bunyaviridae
Colobus polykomos (monkey) leukemia: *see* oncovirus			
Colorado tick fever	type species of subgroup	orbivirus	Reoviridae
Columbia SK (murine encephalomyocarditis)		cardiovirus	Picornaviridae
Columbid herpes h1: *see* pigeon herpes 1			

Congo hemorrhagic fever: *see* Crimean–Congo

Cormorant herpes (h1)   Herpesviridae

CORONAVIRIDAE[42]: Spherical and pleomorphic particles of 60–220 nm diameter (buoyant density in sucrose 1.18 g/cm^3) with characteristic club-shaped widely-spaced about 20 nm projections that give the particle the so-called corona-like appearance. The helical nucleocapsid consists of a plus-strand RNA of about 6×10^6 daltons carrying poly (A) and a phosphoprotein of about 55×10^3 daltons at the 3′ and 5′ end, respectively. Three or four proteins and glycoproteins as well as lipids make up the envelope and, a large glycoprotein the peplomers. The avian infectious bronchitis virus is the type species, and others are human corona virus (HCV), murine hepatitis virus, porcine hemagglutinating encephalitis, and transmissible gastroenteritis virus, as well as probably canine, calf, rat, and turkey coronaviruses. Only three different serotypes have been detected (Figure 4).

Coronavirus enteritis of turkeys: *see* blue comb disease virus.

Corriparta (mosquito-transmitted)   type species of subgroup (3 serotypes)   orbivirus   Reoviridae

Coryza   rhinovirus   Picornaviridae

Cotia[43]: probably separate genus of Entomopoxvirinae (no serological relations detected)

Cottontail herpes (h1)   Gammaherpesvirinae

Cowbane ridge   flavivirus   Togaviridae

Cowpox   orthopox   Chordopoxvirinae

Coxsackie   enterovirus   Picornaviridae

Crane herpes (h1)   Herpesviridae

Virus	Group or subgroup	Genus	Subfamily or family
Crassostrea virginica			Herpesviridae
Crawley (of birds)		orthoreovirus	Reoviridae
Creutzfeldt–Jacob disease[44]	unclassified (probably nonviral)		proposed term: Prion
Cricetid herpes: *see* hamster herpesvirus			
Cricket paralysis[45]	unclassified (possibly strain of encephalomyocarditis)		Picornaviridae
Crimean–Congo hemorrhagic fever	type species of subgroup	nairovirus	Bunyaviridae
C-type particles: *see* Oncovirinae			
Cydia pomonella (coddling moth)	granulosis virus (B)		Baculoviridae
Cypovirus: genus of Reoviridae (type species cytoplasmic polyhedrosis virus of insects).			
Cyprimid herpes: *see* carp poxvirus			
Cytomegalo[46]: genus of Betaherpesvirinae (type species: human cytomegalovirus)			
Cytoplasmic polyhedrosis[47], (Figure 5)	type 1012	cypovirus	Reoviridae
D'Aguilar (*Culicoides*-transmitted)	Palyam	orbivirus	Reoviridae
Dakar bat		flavivirus	Togaviridae
Dane particle: *see* hepatitis B virus			

Name			
Darna trima	nudaurelia β group		
DBS/1 (murine leukemia): see oncovirus			
Delta herpes (monkey) (h7)	simplexvirus		Alphaherpesvirinae
Demodema boranensis	A		Entomopoxvirinae
Denso(nucleosis) (insect hosts)[48]: genus of Parvoviridae			
Dependo (common term adenoassociated viruses): genus of Parvoviridae (all serologically interrelated)			
Dera Ghazi Khan	nairovirus	type species of subgroup	Bunyaviridae
Dermolepida albohirtum	A		Entomopoxvirinae
Dhori	unclassified, possibly orthomyxovirus but tick-borne		
Diatraea	densovirus		Parvoviridae
Dog herpes (h1)			Alphaherpesvirinae
Domestic goat herpes (h2)			Alphaherpesvirinae
Douglas	Simbu	bunyavirus	Bunyaviridae
Drosophila A and P[49]	unclassified (similar to bee slow and acute paralysis virus)		
Drosophila C			Picornaviridae
Drosophila F	(10 ds RNAs)	possibly orthoreovirus	Reoviridae
Drosophila Sigma[49]	possibly birnavirus group		

Virus	Group or subgroup	Genus	Subfamily or family
Drosophila X[49]: possibly birnavirus group			
Duck plaque (h1)			Herpesviridae
Duck spleen necrosis: *see* spleen necrosis virus			
Dugbe		nairovirus	Bunyaviridae
Duo		obsolete name for rotavirus	Reoviridae
Duvenhage (vertebrates)[50]		lyssavirus	Rhabdoviridae
Eastern equine encephalitis		alphavirus	Togaviridae
Ebola: unclassified, very long rods resembling Marburg virus; lethal to man, possibly Rhabdoviridae			
EBV: *see* Epstein–Barr virus			
ECHO 9 (synonym for coxsackie virus A23)		enterovirus	Picornaviridae
ECHO serotype 10, 12, 28[51]		rhinoviruses	Picornaviridae
Ectromelia (mouse)		orthopox	Chordopoxvirinae
ED 1M (mouse)		rotavirus	Reoviridae
Edgehill (mosquito-transmitted)		flavivirus	Togaviridae
Eel: European	similar, but serologically not related, to infectious pancreatic necrosis virus		

Eg An 1825–61	probably uukuvirus	Bunyaviridae
Eggdrop syndrome (chicken)[52]	adenovirus	Adenoviridae
Egtved (synonym: hemorrhagic septicemia of fishes)[53]		probably Rhabdoviridae
EHD (New Jersey, Can Alberta): *see* epizootic disease of deer	orbivirus	Reoviridae
Elapid herpes: *see* Banded Krait, Indian cobra, Siamese cobra		
Elephant herpes (h1)		Herpesviridae
Elephant pox (related to cowpox)	orthopoxvirus	Chordopoxvirinae
Encephalomyocarditis (EMC)[54]	cardiovirus	Picornaviridae
Enchrytraeus fragmentosus (microannelid)	unclassified (270 × 50 nm particles)	
Endogenous *Drosophila* line		possibly Nodaviridae
Engelbreth–Holm avian sarcoma: *see* oncovirus		
Enseada	probably bunyavirus	Bunyaviridae
Entamoeba histolytica (amoeba)	unclassified (40 and 70 nm particles)	
Entebbe bat	flavivirus	Togaviridae
Entero: genus of Picornaviridae (type species: human poliovirus)		
Entomopoxvirinae[55]: subfamily of Poxviridae (of insects)		
Epiphyas postvitana	cotiavirus	Entomopoxvirinae

Virus	Group or subgroup	Genus	Subfamily or family
Epizootic hemorrhagic disease of deer	type species of subgroup (20 serotypes)	orbivirus	Reoviridae
Epstein–Barr[56] (human herpes, h4)	type species	lymphocrytovirus	Gammaherpesvirinae
equ 1 (*Equus caballus*, horse)		mastadenovirus	Adenoviridae
Equid: *see* equine			
Equine abortion (h1)			Alphaherpesvirinae
Equine arteritis[57]		possibly pestivirus	Togaviridae
Equine dependo		dependovirus	Parvoviridae
Equine encephalitis: *see* Eastern and Western equine encephalitis			
Equine encephalosis	(five serotypes)	orbivirus	Reoviridae
Equine herpes h1[58]: *see* equine abortion and rhinopneumonitis virus			
Equine herpes h2[58]: *see* slow-growing cytomegalo-like virus			
Equine herpes h3[58]: *see* coital exanthema virus			
Equine infectious anemia[59]			Lentivirinae
Equine infectious arteritis: *see* equine arteritis			
Equine rhino	(serotypes 1,2)	rhinovirus	Picornaviridae
Equine rhinopneumonitis (h1)		poikilovirus	Alphaherpesvirinae
Equine viral arteritis		alphavirus	Togaviridae
Erythroblastosis: *see* avian erythroblastosis virus			

Name			Family
Erythrocyte aplasia agent			Parvoviridae
Esh: *see* avian sarcoma virus			
Eubenangea	type species of subgroup	orbivirus	Reoviridae
European eel: *see* eel virus			
European ground-squirrel cytomegalo (h1)			Betaherpesvirinae
European swine fever: *see* hog cholera			
European tick-borne encephalitis: *see* tick-borne encephalitis virus			
EV (enterovirus) type 70: *see* acute hemorrhagic conjunctivitis			
EVU 16 (possibly agent of acute infectious lymphocytosis of children)	(unusually large VP4, 49 $\times$ 10^3 dalton)		possibly Picornaviridae
Eyach (tick-transmitted)	Colorado tick fever	orbivirus	Reoviridae
FA: strain of murine encephalomyelitis virus			
Facey's Paddock	Simbu	bunyavirus	Bunyaviridae
Falcon inclusion body disease (h1)			Herpesviridae
Falconid herpes h1: *see* falcon inclusion body disease virus			
Farralon	Hughes	nairovirus	Bunyaviridae
FBJ (murine leukemia): *see* oncovirus			
Feline ataxia		parvovirus	Parvoviridae
Feline calici		calicivirus	Caliciviridae

Virus	Group or subgroup	Genus	Subfamily or family
Feline corona		coronavirus	Coronaviridae
Felid herpes h1: *see* cat herpes virus and infectious rhinotracheitis virus; h2: *see* cat cytomegalovirus.			
Feline infectious peritonitis		possibly coronavirus	Coronaviridae
Feline infectious rhinotracheitis (h1)			Alphaherpesvirinae
Feline leukemia: *see* oncovirus			
Feline panleukopenia		parvovirus	Parvoviridae
Feline parvo: synonym for feline panleukopenia virus			
Feline rhinotracheitis: synonym for feline infectious rhinotracheitis			
Feline sarcoma: *see* oncovirus			
Feline syncytial			Spumavirinae
Fer de Lance[60]		paramyxovirus	Paramyxoviridae
Fibroma (of rabbits, hares, squirrels)	myxoma	leporipoxvirus	Chordopoxvirinae
Figulus sublaevis (*Coleoptera*)		A	Entomopoxviridae
Fijivirus: genus of reoviridae of plants (*see* Section II)			Reoviridae
Filoviridae[61]: proposed family name for the group of Ebola, Marburg, etc. viruses			
Finch paramyxo		paramyxovirus	Paramyxoviridae
Fin isolate (tick-transmitted)	Kemorovo	orbivirus	Reoviridae
Flacherie (of silkworms): *see* Ina-flacherie virus			

Flanders (vertebrates and invertebrates)			probably Rhabdoviridae
Flavi: genus of Togaviridae[62] (type species: yellow fever virus)			
Flexal			probably Arenaviridae
Flock House			possibly Nodaviridae
Flu: *see* Influenza			
Foamy virus group: official term: spumavirinae, subfamily of Retroviridae			Spumavirinae
Foot and mouth disease	type species of	aphthoviruses	Picornaviridae
Fort Morgan		alphavirus	Togaviridae
Fowl adeno 1 (*gal* 1) (CELO)		aviadenovirus	Adenoviridae
Fowl plague		type A	Orthomyxoviridae
Fowlpox		avipoxvirus	Chordopoxvirinae
Fraser Point	Hughes	nairovirus	Bunyaviridae
Friend murine leukemia: *see* oncovirus			
Friend polycythemia induction: *see* oncovirus			
Friend spleen focus-forming: *see* oncovirus			
Frijoles		probably phlebovirus	Bunyaviridae
Frog 3[63]	Singular properties with 100×10^6 dalton DNA, circularly permuted and terminally redundant, resembling phage P22 DNA. Twenty percent of the cytidines are 5-methyl C.		

Virus	Group or subgroup	Genus	Subfamily or family
Frog 4 (h2)			Herpesviridae
Fujinami sarcoma and Fujinami-associated: *see* oncovirus			
G 8886, 15534		orbivirus	Reoviridae
GA: related to SM feline sarcoma virus			
Gabek Forest		phlebovirus	Bunyaviridae
gal 1–9 (*Gallus domesticus*, fowl)		aviadenovirus	Adenoviridae
Galleria mellonella		densovirus	Parvoviridae
Gallid herpes h1: *see* infectious laryngotracheitis			
Gallid herpes h2: *see* Marek's disease			
Gamboa	type species of group	bunyavirus	Bunyaviridae
Gammaherpesvirinae[64]: subfamily of Herpesviridae (lymphocyte-associated)			
Gan Gan	possibly Maputta	bunyavirus	Bunyaviridae
Ganjan	Nairobi sheep disease	nairovirus	Bunyaviridae
Gardner–Arnstein feline leukemia: *see* oncovirus			
Gastroenteritis of man			possibly Parvoviridae
Gazdar murine sarcoma: *see* oncovirus			
GD VII: strain of mouse encephalitis virus			
Geotrupes sylvaticus (*Coleoptera*)		A	Entomopoxvirinae

Germiston	bunyamwera	Bunyaviridae
Getah		Flaviviridae
	alphavirus	
Gibbon ape leukemia: *see* oncovirus		
Goat pox	capripoxvirus	Chordopoxvirinae
Goeldichironomus holoprasimus (*Diptera*)	C	Entomopoxvirinae
Gonometa		possibly Picornaviridae
Goose parvo	parvovirus	Parvoviridae
Gordil	phlebovirus	Bunyaviridae
Gorilla herpes (h3)		Gammaherpesvirinae
Graff: *see* Marburg virus		
Grand Arbout	uukuvirus	Bunyaviridae
Granulosis: subgroup B of Baculoviridae		
Great Island (tick-transmitted)	Kemorovo	orbivirus	Reoviridae
Green iguana (h1)		Herpesviridae
Grey lizard (h1)		Herpesviridae
Grey Lodge		probably Rhabdoviridae
Gross murine leukemia: *see* oncovirus		

Virus	Group or subgroup	Genus	Subfamily or family
Ground squirrel hepatitis[65]: *see* hepatitis B			
GU 71u344	probably Capim	bunyavirus	Bunyaviridae
Guajara	Capim	bunyavirus	Bunyaviridae
Guama	type species of group	bunyavirus	Bunyaviridae
Guaratuba	Guama	bunyavirus	Bunyaviridae
Guaroa	Bunyamwera	bunyavirus	Bunyaviridae
Guinea pig cytomegalo (h2)			Betaherpesvirinae
Guinea pig herpes (h1)			Herpesviridae
Guinea pig sarcoma: *see* oncovirus			
Gumbo Limgo	C	bunyavirus	Bunyaviridae
Gyratrix hermaphroditis (platyhelminth)	unclassified (70 nm particles)[66]		
H1, 2, 3 (hamster and rat osteolytic viruses)[66]		parvovirus	Parvoviridae
h1 to 34 types of human adeno		mastadenovirus	Adenoviridae
H 32580	Anopheles A	bunyavirus	Bunyaviridae
Haden: *see* hemadsorbing enteric virus			
Hamster herpes (h1)			Betaherpesvirinae
Hamster papillomavirus		papillomavirus	Papovaviridae

Hantaan (of rodents)	(etiological agent of Korean hemorrhagic fever with renal symptoms) related to Prospect Hill virus	possibly new genus	Bunyaviridae
HAPV (hamster polyomavirus)		polyomavirus	Papovaviridae
Hare fibroma		leporivirus	Chordopoxvirinae
Harris strain: *see* oncovirus			
Hartebeest herpes (h2)			Gammaherpesvirinae
Hart Park (vertebrates and invertebrates)			probably Rhabdoviridae
Harvey sarcoma (murine): *see* oncovirus			
Hazara[67]		nairovirus	Bunyaviridae
HBI: a variant of MC 29			
HD[68]	human (single isolate)	polyoma	Papovaviridae
HEL-12: *see* oncovirus			
Heliothis armigeras		cotiavirus	Entomopoxvirinae
Hemadsorbing enteric virus (bovine), (HADEN)		parvovirus	Parvoviridae
Hemadsorption virus	types 1 and 2 parainfluenza	paramyxovirus	Paramyxoviridae

Virus	Group or subgroup	Genus	Subfamily or family
Hemagglutinating encephalomyelitis (of pigs)[69]		coronavirus	Coronaviridae
Hemorrhagic conjunctivitis: *see* acute hemorrhagic conjunctivitis			
Hemorrhagic encephalitis of Japan: *see* Sendaivirus			
Hemorrhagic encephalopathy (rats)		parvovirus	Parvoviridae
Hemorrhagic fever		arenavirus	Arenaviridae
Hemorrhagic septicemia (salmon): possibly identical to Egtvedvirus			probably Rhabdoviridae
Hepatitis A (old term, infectious hepatitis)[70,71]	human (very restricted host range)	enterovirus	Picornaviridae

Hepatitis B (old term, serum hepatitis)[71,72]: This virus was regarded as a virus infecting only man until recently when very similar and probably related viruses were discovered in woodchucks, Pekin ducks, and ground squirrels. The virus occurs in the blood as the "Dane" particle of 42 nm diameter, the core of 22 nm diameter, and the DNA (molecular weight of about 2×10^6) consisting of a double-stranded circular molecule, one strand of 3200 nucleotides, and the other, the plus strand being incomplete. The DNA polymerase of the virion fills the gap bringing the molecular weight to 2.1×10^6. Replication involves RNA–DNA hybrid forms, thus resembling Retroviridae. Besides this protein, there is the surface protein of about 22×10^3 daltons, the core protein of about 18×10^3 daltons, and possibly another antigenic component derived from the others by aggregation. The virus is readily transmitted by various routes, extremely widespread, not always markedly pathogenic, and frequently latent and persistent. It represents now one of the main epidemiological problems of man.

Virus	Group or subgroup	Genus	Subfamily or family
Hepatosplenitis of owl	strigid	herpesvirus h1	Herpesviridae
Herpes aotus	aotine	herpesvirus h2	Gammaherpesvirinae
Herpes aotus (owl monkey)	aotine	herpesvirus h3	Betaherpesvirinae

Herpes ateles strain 73, 810 (spider monkey)[73] (h2, h3)	rhadinovirus	Gammaherpesvirinae
Herpes B (h2)	simplexvirus	Alphaherpesvirinae
Herpes cuniculi (cottontail rabbit) (h2)		Herpesviridae
Herpes cyclopsis (monkey) (h3)		Herpesviridae
Herpes M (h1)		Alphaherpesvirinae
Herpes pan (chimpanzee) (h1)	lymphocrytovirus	Gammaherpesvirinae
Herpes papio (monkey) (h12)	simplexvirus	Gammaherpesvirinae
Herpes platyrrhinae (monkey) (h2)		Gammaherpesvirinae
Herpes pottos (kinkajou) (h2)		Herpesviridae
Herpes saimiri (monkey)[73] (h2)	rhadinovirus	Gammaherpesvirinae
Herpes salmonis (salmon) (h1)		Herpesviridae
Herpes sanguinus (marmoset) (h1)	simplexvirus	Herpesviridae
Herpes scophthalmus (turbot) (h1)		Herpesviridae
Herpes simplex (human) types 1, 2 (h1, h2)	simplexvirus	Alphaherpesvirinae
Herpes sylvagus (cottontail rabbit) (h1)		Gammaherpesvirinae
Herpes T (monkey) (h1)		Alphaherpesvirinae
Herpes tamarinus (monkey) (h1)		Alphaherpesvirinae

| Virus | Group or subgroup | Genus | Subfamily or family |

HERPESVIRIDAE[74]: More than 80 herpes viruses are known that were isolated from many different hosts. The virion of 120–200 nm (density in CsCl about 1.25 g/cm^3) consists of a nucleocapsid core, an icosahedral shell (105 nm diameter) of 162 capsomers, the tegument, and the envelope with surface projections (Figure 6). The molecular weight of the linear double-stranded DNA ranges from 80–150 × 10^6 for different genera, and that of the more than 20 proteins from 12–200 × 10^3. Several proteins are phosphorylated. The envelope contains lipid and glycoproteins. Three subfamilies have been established differing in biological properties. The DNA of ALPHAHERPESVIRINAE is usually less than 100 × 10^6 daltons. These viruses have wide host ranges, a short reproductive cycle, rapid cytopathology, and frequently cause latent infection of ganglia.

The BETAHERPESVIRINAE have DNAs near 150 × 10^6 daltons. They have a narrow host range, a long reproductive cycle, slowly progressing cytopathology frequently causing cell enlargement (cytomegaly), and late infections can become established in many different tissues.

The GAMMAHERPESVIRINAE (DNA molecular weights usually near 100 × 10^6) have narrow in vivo host ranges. They infect specifically either B or T lymphocytes, either in lytic or in persistent manner. Thus, the reproductive cycle varies in length and ensuing cytopathology, with frequent latent infections.

Proposed genera for the ALPHAHERPESVIRINAE are SIMPLEXVIRUS (herpes simplex-like viruses); species human herpesvirus 1 and 2, bovine herpesvirus 2, cercophithecine herpesvirus 1 and 2 with common names herpes simplex 1, 2, bovine mammilitis, SA8 and B virus; POIKILOVIRUS (pseudorabies-like viruses): species suid herpes virus 1 or pseudorabies, and equid herpesvirus 1 or equine rhinopneumonitis virus; and VARICELLAVIRUS (varicella–Zoster–like viruses): species human herpesvirus 3 or varicella–Zoster virus. The proposed BETAHERPESVIRINAE genera are CYTOMEGALOVIRUS: human herpesvirus 5 or cytomegalovirus, and MUROMEGALOVIRUS: murid herpesvirus 5 or cytomegalovirus. For the GAMMAHERPESVIRINAE, proposed genera are LYMPHOCRYTOVIRUS (Epstein–Barr–like viruses): human herpesvirus 4 (or Epstein–Barr virus), cercopithecine herpes virus 12 or baboon herpesvirus, pongine herpesvirus 1 or chimpanzee herpes virus; THETALYMPHOCRYPTOVIRUS: Marek's disease virus and melleagrid herpesvirus 1 (turkey herpes virus); and RHADINOVIRUS (saimiri–ateles-like herpes viruses): ateline herpesvirus 2 and 3 or herpesvirus ateles, strains 810 and 73, and saimiriine herpesvirus 2 or herpesvirus saimiri (Figure 6).

Heteronychos arator: see blackbeetle virus

Hog cholera (identical to swine fever)		pestivirus	Togaviridae
Housefly[75]	similarities to (10 ds RNAs) and differences from others	orthoreovirus	Reoviridae
Hsuing–Kaplow (h1)			Herpesviridae
HT		parvovirus	Parvoviridae
HTLV: *see* Human T-cell leukemia			
Huacho (tick-transmitted)	Kemorovo	orbivirus	Reoviridae
Hughes	type species of group	nairovirus	Bunyaviridae
Human 72: *see* hepatitis A virus			
Human adenoviruses		adenovirus	Adenoviridae
Human corona (identical with enteric virus, respiratory virus)		coronavirus	Coronaviridae
Human coxsackie	A1–22, A24, B1–B6	enterovirus	Picornaviridae
Human cytomegalo (h5)	type species	cytomegalovirus	Betaherpesvirinae
Human echo	type 1–9, 11–27, 29–34 (9 = entero A23)	enterovirus	Picornaviridae
Human enteric		coronavirus	Coronaviridae
Human entero	type 68–71	enterovirus	Picornaviridae
Human foamy			Spumavirinae
Human herpes (simplex) type 1		simplexvirus	Alphaherpesvirinae

Virus	Group or subgroup	Genus	Subfamily or family
Human herpes (simplex) type 2		simplexvirus	Alphaherpesvirinae
Human herpes type 3 (h3) (varicella–zoster virus)		varicellavirus	Alphaherpesvirinae
Human herpes type 4 (h4) (Epstein–Barr virus)		lymphocrytovirus	Gammaherpesvirinae
Human herpes type 5 (h5)		cytomegalovirus	Betaherpesvirinae
Human infantile enteritis (or gastroenteritis)		rotavirus	Reoviridae
Human papilloma (wart)	type 1a	papillomavirus	Papovaviridae
Human respiratory		coronavirus	Coronaviridae
Human rhino	1A-type species, 2–113	rhinovirus	Picornaviridae
Human rota: *see* infantile (gastro)enteritis			
Human T-cell leukemia[76] (HTLV I, II)		oncovirus, possibly type D	Retroviridae
HTLV III: possibly responsible for AIDS (acquired immune deficiency syndrome); related to similar simian virus		possibly visnavirus	Retroviridae
Hydra vulgaris (coelenterate)		adenovirus	Adenoviridae
HZ-1	(nonoccluded) (D) (two envelopes) (nonoccluded) (C)		
Ib Ar 22619, 33853	epizootic disease of deer subgroup	orbivirus	Reoviridae

Ibaraki (cattle)	epizootic disease of deer	orbivirus	Reoviridae
IBH 11306, 13019 (*culicoides*-transmitted)	Palyam	orbivirus	Reoviridae
IBV: *see* avian infectious bronchitis virus			
Ictalurid h1: *see* channel catfish herpes			
Icoaraci		phlebovirus	Bunyaviridae
Icosahedral cytoplasmic DNA viruses (ICDV): obsolete term for viruses now classified as Iridoviridae			
Ife		orbivirus	Reoviridae
Ignotosoma sabellarium (ciliate) unclassified (75 × 50 nm particles)			
Iguanid herpes: *see* Green Iguana herpes h1			
Ilesha	Bunyamwera	bunyavirus	Bunyaviridae
Ilhéus (mosquito-borne)		flavivirus	Togaviridae
IMC-Hz-1-NOV – *Heliothis zea*[77]	nuclear polyhedrosis (A)		Baculoviridae
IM: possibly responsible for multiple sclerosis			
Inachis io	type 2	cypovirus	Reoviridae
Ina-Flacherie		probably dependovirus	Parvoviridae
Inclusion body rhinitis (swine) (h2)			Betaherpesvirinae
Indian cobra herpes (h1)			Herpesviridae
Indiana: strain of vesicular stomatitis virus			

Virus	Group or subgroup	Genus	Subfamily or family
Infantile (gastro)enteritis (many animals, including humans)		rotavirus	Reoviridae
Infectious anemia (horse): see equine infectious anemia			
Infectious arteritis (horse): see equine arteritis			
Infectious bovine keratoconjunctivitis			Herpesviridae
Infectious bovine rhinotracheitis (h1)		simplexvirus	Alphaherpesvirinae
Infectious bronchitis (chickens), (IBV)		coronavirus	Coronaviridae
Infectious bursal disease (chicken)		possibly Birnavirus group	
Infectious canine hepatitis		mastadenovirus	Adenoviridae
Infectious canine laryngotracheitis		mastadenovirus	Adenoviridae
Infectious hematopoietic necrosis			probably Rhabdoviridae
Infectious laryngotracheitis (chicken) (h1)			Alphaherpesvirinae
Infectious pancreatic necrosis (fish): type species of proposed birnavirus group			
Infectious pustular vulvovaginitis (cattle) (h1)			Alphaherpesvirinae
Infectious rhinotracheitis (cat) (h1)			Alphaherpesvirinae
Influenza[78] (Figure 7)		type A, B, C	Orthomyxoviridae

Ingwavuma		Simbu	bunyavirus	Bunyaviridae
Inini		Simbu	bunyavirus	Bunyaviridae
Inkoo		California	bunyavirus	Bunyaviridae

Insect viruses: *see* Baculoviridae, Iridoviridae, Parvoviridae, Poxviridae, etc.

| Iridescent viruses, type 1–20: different insect hosts | | | iridovirus | Iridoviridae |

IRIDOVIRIDAE[79]. A rather large heterogeneous group of large icosahedral particles (130–300 nm diameter, 1300–4500 S, density 1.16–1.35 g/cm^3). They contain one or two molecules of double-stranded linear DNA of $100-250 \times 10^6$ daltons. The small and large iridescent insect viruses, genus IRIDOVIRUS and CHLORIRIDOVIRUS, respectively, are more obviously interrelated than the frog viruses (genus RANAVIRUS) and the not yet classified African swine fever and lymphocystic disease virus groups. The unenveloped iridescent viruses contain, besides the DNA, 15–20 proteins (about 75%) and lipid (about 6%). The chloriridoviruses have envelopes derived from the plasma membrane, thus represent the larger, ether-sensitive lipid-rich iridoviridae. They have narrow host ranges. These and the iridoviruses (type species Tipula iridescent virus) reach very high concentrations in their insect hosts, thus turning the moribund larvae irridescent. African swine fever replicates in and is transmitted by ticks (Figure 8).

Irituia		Changuinola	orbivirus	Reoviridae
Isfahan: strain of vesicular stomatitis virus				
Israel turkey meningitis			flavivirus	Togaviridae
Ita			orbivirus	Reoviridae
Itaituba			phlebovirus	Bunyaviridae
Itaporanga			phlebovirus	Bunyaviridae
Itaqui	C		bunyavirus	Bunyaviridae

Virus	Group or subgroup	Genus	Subfamily or family
Itimirim	Guama	bunyavirus	Bunyaviridae
J (mouse)		paramyxovirus	Paramyxoviridae
Jacareacanga	Corriparta	orbivirus	Reoviridae
Jamestown Canyon	California	bunyavirus	Bunyaviridae
Japanant		orbivirus	Reoviridae
Japanese encephalitis[80]		flavivirus	Togaviridae
JC (human)[81] (agent responsible for progressive multifocal leukoencephalopathy)		polyomavirus	Papovaviridae
Jerry Slough	California	bunyavirus	Bunyaviridae
JHM (murine)[82]		coronavirus	Coronaviridae
Joagsiekt: *see* sheep pulmonary adenomatosis virus			
Joinjakaka (invertebrate hosts)			probably Rhabdoviridae
Juan Diaz	Capim	bunyavirus	Bunyaviridae
Jugra (mosquito-transmitted)		flavivirus	Togaviridae
Junco		avipoxvirus	Chordopoxvirinae
Junin (human pathogen)	Takaribe complex	arenavirus	Arenaviridae
Junonia		densovirus	Parvoviridae
Jurona		bunyavirus	Bunyaviridae

Jutiapa		flavivirus	Togaviridae
K (mouse)		polyomavirus	Papovaviridae
Kadam (mosquito-transmitted)		flavivirus	Togaviridae
Kaeng Khoi		probably bunyavirus	Bunyaviridae
Kaikalur	Simbu	bunyavirus	Bunyaviridae
Kairi	Bunyamwera	bunyavirus	Bunyaviridae
Kaisodi	type species of possible group	bunyavirus	Bunyaviridae
Kamese (invertebrate hosts)			probably Rhabdoviridae
Kao Shuan	Dera Ghazi Khan	nairovirus	Bunyaviridae
Karimabad	Sandfly fever	phlebovirus	Bunyaviridae
Karshi (tick-borne)		flavivirus	Togaviridae
Kasba	Palyam	orbivirus	Reoviridae
Kawino (host, mosquito *Mansonia uniformis*)[83]		probably enterovirus (though no poly (A))	Picornaviridae
Kedongon (tick-borne)		flavivirus	Togaviridae
Kelp fly[84]	unclassified (29 nm particles, 158 S, RNA 3.5×10^6, proteins 73 and 29 $\times 10^3$ daltons)		
Kemerovo	type species (20 serotypes)	orbivirus	Reoviridae

Virus	Group or subgroup	Genus	Subfamily or family
Kenai (tick-transmitted)	Kemorovo	orbivirus	Reoviridae
Kern Canyon (of bats)			probably Rhabdoviridae
Ketapang	possibly Bakau	bunyavirus	Bunyaviridae
Keuraliba (vertebrate hosts)			probably Rhabdoviridae
Keystone	California	bunyavirus	Bunyaviridae
Khasan		possibly bunyavirus	Bunyaviridae
Kilham (rat)	type species	parvovirus	Parvoviridae
Kimberley (invertebrate hosts)			probably Rhabdoviridae
Kinkajou (h1)			Herpesviridae
Kirk (rat)		parvovirus	Parvoviridae
Kirsten (murine): *see* oncovirus			
Kissling			Picornaviridae
Klamath (mouse)		cardiovirus	probably Rhabdoviridae
Kokobera (mosquito-borne)		flavivirus	Togaviridae
Koongol	type species of group	bunyavirus	Bunyaviridae

Name		Genus	Family
Kotankan (transmitted by midges)		lyssavirus	Rhabdoviridae
Koutango		flavivirus	Togaviridae
Kowanyama		bunyavirus	Bunyaviridae
Kunjin (mosquito-borne)[85]		flavivirus	Togaviridae
Kununurra (invertebrates)			probably Rhabdoviridae
Kuru	unclassified (probably nonviral)		proposed term: Prion
Kwatta (invertebrates)			probably Rhabdoviridae
Kyasanur Forest disease (tick-borne)		flavivirus	Togaviridae
Kyzylagach		alphavirus	Togaviridae
Lacertid: see Green Lizard herpes			
La Crosse[86]	California	bunyavirus	Bunyaviridae
Lactic dehydrogenase (mice)		possibly pestivirus	Togaviridae
Lagos bat		lyssavirus	Rhabdoviridae
Lake Victoria (cormorant)			Herpesviridae
Langat (tick-borne)		flavivirus	Togaviridae
Lanjan	Kaisodi	possibly uukuvirus	Bunyaviridae
Lapine		parvovirus	Parvoviridae

Virus	Group or subgroup	Genus	Subfamily or family
Laryngotracheitis (canine)			Adenoviridae
Laspeyresia pomonella		cotiavirus	Entomopoxvirinae
Lassa (human pathogen)		arenavirus	Arenaviridae
Latino	Tacaribe complex	arenavirus	Arenaviridae
LAV (leukoadenopathy virus): possibly identical with HTLV III			
LCM group: *see* lymphocytic choriomeningitis			
Leafhopper A		cypovirus	Reoviridae
Lebombo (mosquito-borne), (infects man)		orbivirus	Reoviridae
Lednice	Turlock	bunyavirus	Bunyaviridae
Lentivirinae:[87] (so-called slow viruses) subfamily of Retroviridae			
Leporid herpes h1: *see* cottontail herpes virus, herpes sylvilagus			
Leporid herpes h2: *see* herpes cunciculi			
Leporipox (rabbit myxoma): genus of Chordopoxvirinae			
Leukoviruses: obsolete term for oncovirinae			
Lipovnik (tick-borne)	Kemorovo	orbivirus	Reoviridae
Liverpool vervet monkey (h6)			Alphaherpesvirinae
LK (horse)			Herpesviridae
Llano seco		orbivirus	Reoviridae

LNV$_G$ (*Lymantria* spp.)	(lower molecular weights, but slight serological relationship with black beetle virus)		probably Nodaviridae
Lokern	Bunyamwera	bunyavirus	Bunyaviridae
Lone Star		bunyavirus	Bunyaviridae
Lorisine herpes h1: *see* Kinkajou virus, herpes pottos			
Louping Ill (tick-borne)		flavivirus	Togaviridae
L-S (rat)		parvovirus	Parvoviridae
LT 1–4 (newt)		ranavirus	Iridoviridae
LU III		parvovirus	Parvoviridae
Lucké (frog adenocarcinoma) (h1)			Herpesviridae
Lukuni	Anopheles A	bunyavirus	Bunyaviridae
Lumbo	probably California	bunyavirus	Bunyaviridae
Lumpy skin disease (Neethling)		capripox	Chordopoxvirinae
Lymantria dispar	nuclear polyhedrosis virus (A) (licensed for use as insecticide)		Baculoviridae

Lymphocryto: genus of Gammaherpesviridae (type species: Epstein–Barr virus)

Virus	Group or subgroup	Genus	Subfamily or family
Lymphocystis disease (of fishes): possible genus of Iridoviridae (Figure 9); 200 nm diameter icosahedra			
Lymphocytic choriomeningitis[88]	type species		Arenaviridae
Lymphoproliferative group: old term for rhadino and lymphocrytovirus genera of Gammaherpesvirinae			
Lymphotropic polyoma (LPV) of green monkey (and man?)		papovavirus	Papovaviridae
Lyssa[89]: genus of Rhabdoviridae (type species: rabiesvirus)			
M9		orbivirus	Reoviridae
M25	type 4	parainfluenza	Paramyxoviridae
MAC-1 (*Macaca arctoides* and *mulata*) leukemia: *see* oncovirus			
Machupo (human pathogen)	Tacaribe complex	arenavirus	Arenaviridae
Macropipius depurator (crab) unclassified (55 and 150–300 nm particles)			
Macropodid herpes h1: *see* Parma wallaby herpes virus			
Madrid	C	bunyavirus	Bunyaviridae
Maedi (so-called slow virus)			Lentivirinae
Magwari	Bunyamwera	bunyavirus	Bunyaviridae
Mahogany Hammock	Guama	bunyavirus	Bunyaviridae
Main Drain	Bunyamwera	bunyavirus	Bunyaviridae

Malignant Catarrhal fever of wildebeest (also cattle, deer, etc. in Africa) (h1)			probably Gammaherpesvirinae
Malignant Catarrhal fever of wildebeest (in America)		cytomegalovirus	Betaherpesvirinae
Mammalian adenovirus: *see mas*			
Mammalian C type or type C: *see* oncovirus			
Mammary tumor: *see* mouse mammary tumor virus			
Manawa		uukuvirus	Bunyaviridae
Manzanilla	Simbu	bunyavirus	Bunyaviridae
Mapputta	type species of group		possibly Bunyaviridae
Maprik	Maputta		possibly Bunyaviridae
Marburg (monkey, man)	unclassified (variably long, 130–2000, average 665 × 80 nm rods), lethal in man		possibly Rhabdoviridae proposed: Filoviridae
Marco (vertebrate hosts)			probably Rhabdoviridae
Marek's disease (chicken)[90] (h2)		thetalymphocrytovirus	Gammaherpesvirinae
Marituba	C	bunyavirus	Bunyaviridae
Marmoset cytomegalo		cytomegalovirus	Betaherpesvirinae

Virus	Group or subgroup	Genus	Subfamily or family
Mason–Pfizer (monkey)	type D	oncovirus	Oncovirinae
Mastadenovirus[91]: genus of mammalian Adenoviridae			
Matruh	Tete	bunyavirus	Bunyaviridae
Maus–Elberfeld (ME): serologically identical with encephalomyocarditis virus			
Mayaro		alphavirus	Togaviridae
MC 29 (avian myelocytomatosis): see oncovirus			
MCF (murine): see oncoviruses			
Measles	type species (RNA of 4.5×10^6 daltons)	morbillivirus	Paramyxoviridae
Medical Lake Macaque herpes (h9)			Alphaherpesvirinae
mel 1, 2 (*Meleagris gallopapo*, turkey)		aviadenovirus	Adenoviridae
Melanoplus sanguinipes (*Orthoptera*)		B	Entomopoxvirinae
Melao	California	bunyavirus	Bunyaviridae
Meleagrid h1: see turkey herpes virus h1			
Melolontha melolontha (*Coleoptera*)	type species	A	Entomopoxvirinae
Mengo[92] (rodents): synonym for encephalomyocarditis			

Mermet	Simbu	bunyavirus	Bunyaviridae
MH2: *see* avian oncovirus (member of the MC29 group)			
Middleburg		alphavirus	Togaviridae
Milker's node		parapoxvirus	Chordopoxvirinae
Minatitlan	type species of group	bunyavirus	Bunyaviridae
Mink (cytopathic or cell focus-inducing or forming, MCF)[93]: *see* spleen focus-inducing			
Mink enteritis		parvovirus	Parvoviridae
Minute virus of mice (MVM)[94]	(two structural and two nonstructural proteins)	parvovirus	Parvoviridae
Mirim	Guama	bunyavirus	Bunyaviridae
Mitchell River (*culicoides*)		orbivirus	Reoviridae
MM C-1 (macaque) leukosis: *see* oncovirus			
MM (hamster): possibly identical to (murine) encephalomyocarditis			
MM strain of BK virus involved in human malignancies, though slightly smaller			
MNV-1 (*Macaca nemestrina*): *see* oncovirus			
Modoc		flavivirus	Togaviridae
Moju	Guama	bunyavirus	Bunyaviridae
Mokola (shrew)		lyssavirus	Rhabdoviridae

Virus	Group or subgroup	Genus	Subfamily or family
Molluscum contagiosum		orthopoxvirus	Chordopoxvirinae
Moloney (murine): *see* oncoviruses			
Monkey cytomegalo (h8)			Betaherpesvirinae
Monkey pox		orthopox	Chordopoxvirinae
Mono Lake	Kemorovo	orbivirus	Reoviridae
Montana myotis leukoencephalitis		flavivirus	Togaviridae
Morbilli[95]: genus of Paramyxoviridae (type species: measles virus)			
Moriche	Capim	bunyavirus	Bunyaviridae
Mosquito iridescent		chloriridovirus	Iridoviridae
Mossuril (vertebrates and invertebrates)			probably Rhabdoviridae
Mount Elgon bat (mosquito-transmitted)			probably Rhabdoviridae
Mouse (*see also* murine) cytomegalo (h1)		muromegalovirus	Betaherpesvirinae
Mouse encephalomyelitis		possibly cardiovirus	Picornaviridae
Mouse hepatitis (MHV)[96]			Coronaviridae
Mouse mammary tumor[97] (Bittner virus)	type B	oncovirus	Oncovirinae
Mouse polio		enterovirus	Picornaviridae

Virus	Group or subgroup	Genus	Subfamily or family
Murine viruses: *see also* mouse viruses			
Murocytomegalo: genus of Betaherpesviridae (type species: murine cytomegalovirus)			
Murray Valley encephalitis (mosquito-borne)		flavivirus	Togaviridae
Murutucu	C	bunyavirus	Bunyaviridae
Murweh: related to Barmah Forest virus			Togaviridae
mus 1 (*Mus musculus*, mouse): *see* minute virus of mice		mastadenovirus	Adenoviridae
MVM: *see* minute virus of mice			
Myelocytomatosis: *see* MC 29			
Mykenes (tick-transmitted)	Kemorovo	orbivirus	Reoviridae
Myxoma		leporipox	Chordopoxvirinae
Naegleria gruberi (amoeba)	unclassified (100 nm particles)		
Nairobi sheep disease	type species of group	nairovirus	Bunyaviridae
Nairovirus[99]: genus of Bunyaviridae (type species: Nairobi sheep disease virus)			
Nariva (murine)		paramyxovirus	Paramyxoviridae
Navarro (vertebrate hosts)			probably Rhabdoviridae
N2B-10-1: *see* oncovirus			

Name			Family
Ndumu		alphavirus	Togaviridae
NE (agent of *Nephropathia epidemica* of rodents)			possibly Bunyaviridae
Nebraska calf diarrhea		rotavirus	Rhabdoviridae
Negishi		flavivirus	Togaviridae
Nelson Bay (avian)		orthoreovirus	Reoviridae
Neonatal calf diarrhea		coronavirus	Coronaviridae
Nepuyo	C	bunyavirus	Bunyaviridae
Newcastle disease	type species	paramyxovirus	Paramyxoviridae
New Jersey: strain of vesicular stomatitis virus			
New Minto (invertebrate hosts)			probably Rhabdoviridae
Newt LT 1–4, T 6–20		ranavirus	Iridoviridae
Nique		phlebovirus	Bunyaviridae
Nodamura[100]		nodavirus	Nodaviridae

NODAVIRIDAE[101]: Isometric 29-nm-diameter particles (135 S, density 1.34 g/cm^3, stable to pH 3) consisting of two plus-strand RNAs, lacking poly(A), of about 1.15 and 0.46×10^6 daltons, both required for infectivity, and an intracellular RNA of 0.15×10^6 daltons, derived from the large RNA. The 104×10^3 dalton gene product is probably the RNA polymerase, that of 46×10^3 daltons the coat protein precursor. Wide host range among insects. Transmissible to suckling mice by *Aedes aegypti*. Most are serologically interrelated. Members: Nodamura, black beetle, Arkansas bee, Boolarra, endogenous Drosophila line, and Flockhouse viruses.

Name			Family
Nola	Simbu	bunyavirus	Bunyaviridae

Virus	Group or subgroup	Genus	Subfamily or family
Nonoccluded enveloped nuclear insect viruses: subgroup C and D of Baculoviridae			
Northway	bunyamwera	bunyavirus	Bunyaviridae
Norwalk		calicivirus	Caliciviridae
Norwalk agent: *see* gastroenteritis of man			
Ntaya (mosquito-borne)		flavivirus	Togaviridae
Nuclear polyhedrosis: subgroup A of Baculoviridae			
Nudaurelia β group[102]: isometric 35-nm-diameter particles (200 S, density in CsCl 1.29 g/cm³, stable at pH 3) consisting of one molecule of plus-strand RNA of 1.8×10^6 daltons. Most members are serologically interrelated. Hosts are *Lepidoptera*. The type species was isolated from *Nudaurelia cytherea capensis*, others from *Antheraea eucalypti, Darna trima, Thosea asigna, Philosamia ricini,* and *Trichoplusia ni.*			
Nugget (tick-transmitted)	Kemorovo	orbivirus	Reoviridae
Nyabira	Palyam	orbivirus	Reoviridae
Nymphalidae spp.		densovirus	Parvoviridae
NZB-10-1 (murine): *see* oncoviruses			
O-agent (sheep, cattle)		rotavirus	Reoviridae
Obodhiang (mosquito-transmitted)		lyssavirus	Rhabdoviridae
OC 43			Coronaviridae
Oceanside		uukuvirus	Bunyaviridae
Octopus vulgaris disease			possibly Iridoviridae

Oita 293			probably Rhabdoviridae
OK 10 (avian acute leukemia): see oncovirus (member of MC 29 group)			
Okhotskiy (tick-transmitted)	Kemorovo	orbivirus	Reoviridae
Olifantsvlei	type species of group	bunyavirus	Bunyaviridae
Omsk hemorrhagic fever		flavivirus	Togaviridae
Oncorynchus masau (salmon) herpes (h12)			Herpesviridae
Oncovirinae (RNA tumor viruses)[103]		subfamily of Retroviridae (*see* oncovirus)	

ONCOVIRUS: On morphological grounds, this genus has been classified as consisting of four subgenera, termed A-type, B-type, C-type, and D-type. The A-TYPE particles, characterized by a toroidal nucleoid, may be developmental stages leading to B-type or D-type virions, or they may be abortive. Thus, they have lost their class status. The B-TYPE virions have a spherical nucleoid that is eccentrically located and they bud as spherical particles with long spikes. The (Bittner) mouse mammary tumor virus and its strains or close relatives seem to be the only representatives of that subgenus. The D-TYPE particles resemble the B-type in buoyant density (1.21 g/cm³ in sucrose) and in budding of complete short surface knobs instead of long spikes. This type also has one well-characterized representative, the Mason–Pfizer monkey virus, isolated from a rhesus monkey's breast carcinoma, but the recently discovered human HTLV viruses may also be D-type viruses.

The C-TYPE particles form only upon budding as crescents, but are virions with centrically located nucleoids and short spikes. Their buoyant density is lower (1.16 g/cm³ in sucrose). They are, in contrast to the other types, extremely numerous, and have resisted all attempts at consistent and generally acceptable classification. The classical "RNA tumor viruses" were given the names either of the person that first described them, or of the host–victim and the disease that seemed to characterize them. But is has become evident that (1) one investigator can discover more than one virus, (2) most viruses can infect more than one species, and (3) the same virus can cause different malignancies in the same or different species, depending on the circumstances. Thus, letter–number combinations are now generally

Virus	Group or subgroup	Genus	Subfamily or family

used in naming new C-type viruses or new variants of the classical ones. Only a classification of the C-type viruses into avian, murine, feline, etc. has become generally accepted, although instances of crossing over between groups are also recognized.

What has complicated the classification of the oncoviruses greatly are the following two facts: Almost all C-type viruses, and possibly all the murine ones, are defective, lacking either the transforming, i.e., oncogenic gene (onc) (transformation-defective, td) or the ability to replicate (replication-defective, rd) by lacking part or all of the env, gag, or pol genes (envelope proteins, group-specific antigen = core protein, or polymerase). In all these instances, only the association with, or presence in the cell of, another oncovirus of complementary deficiency can result in replication and transformation, i.e., oncogenicity. Further, homologues of the transforming gene of the RNA tumor viruses, or at least similar nucleotide sequences, occur in the host genomes, and the same is true for other oncovirus genes and even nonhost animal genomes. Thus, the possibility of the intracellular formation of fully competent viruses and/or oncogenic capability has been assumed to be a likely occurrence.

The preferred classification within each group of C-type viruses is based on their ability to interfere with infection of closely related viruses. Host range is an additional classifying criterion. Both, but particularly interference, represent expression of the nature of their envelope proteins which must interact with specific host cell receptor site structures. However, other classifications have also been proposed.

The AVIAN C-TYPE viruses were the first to be discovered and studied, largely because several strains of Rous SARCOMA virus are nondefective. Five subgroups (A–E) have been identified, each comprising both sarcoma viruses and transformation-defective LEUKOSIS VIRUSES. Subgroup A: Rous sarcoma (RSV)-29, Schmidt–Ruppin RSV1, Prague RSV-A, and Engelbreth–Holm RSV; leukosis virus Rous-associated (RAV)$^{-1, -3, -4,}$ $_{and -5,}$ Fujinami-associated (FAV), and myeloblastosis-associated viruses (MAV), resistance-inducing factor (RIF), and RPL-12. Subgroup B: Schmidt–Ruppin RSV2, Prague RSV-B, and Harris RSV; leukosis viruses RAV-2, RAV-6, MAV-2, and erythroblastosis virus 4. Subgroup C: Prague RSV-C, B77 (Bratislava) and Mill Hill 2; leukosis viruses RAV-7 and RAV-49. Subgroup D: Schmidt–Ruppin RSV-D and Carr–Zilber RSV-D; leukosis RAV-50 and Carr–Zilber-associated virus. Subgroup E: Schmidt–Ruppin RSV-E, Prague RSV-E, and leukosis viruses RAV-0, RAV-60, and induced leukosis viruses. Not yet classified are the Bryan high and standard titer strains of RSV and the Fujinami sarcoma virus; among leukosis viruses, avian myeloblastosis, myelocytoma MC29, and many others.

The leukosis viruses, also called leukosis–leukemia viruses, that lack *onc* genes, can be symptomless, but they can also, depending on conditions and hosts, cause occasionally or regularly severe diseases of the lymphatic system, so-called nonsolid tumors. They become most evident as "helpers" of the replication-defective sarcoma viruses.

The most studied MAMMALIAN C-type viruses are those of rodents, mostly mice and cats. All murine sarcoma and leukemia viruses appear to be defective, and thus always represent mixtures or recombinants. For instance, the Rauscher virus represents a complex of three replication-competent and one defective virus. The latter is responsible for severe erythroblastosis, and is called spleen focus-forming virus (SFFV). Many rodent oncoviruses have been grouped according to their host range in four classes as follows. Among the *ecotropic* viruses, i.e., those characterized by infecting only rodent cells, are the Friend, Rauscher, Moloney, AKR-L, Cas-Br-M, and AKv virus complexes; *amphotropic*, i.e., infecting also many nonrodent cells, are 1504A and 4070A; *xenotropic*, i.e., unable to infect mouse cells, are Balb/2 and Balb-10-1, ATS-124, N2B-10-1, and Cas Eno 1-X. MCF, mink cell focus-forming viruses, represent a fourth group also termed ecoxenotropic or dualtropic.

The feline sarcoma viruses are generally in the same manner defective and helper-dependent as the murine. The replication-competent feline leukemia viruses can act as helper also of murine sarcoma virus in mouse cells. Three subgroups (A, B, and C) have been defined, again on the basis of interference and host range. Most isolates are, however, mixtures of subgroup A and B viruses. For human oncoviruses see HTLV.

Many viruses listed in the catalogue have not yet been classified, many others fail to be listed, and many more arise currently or become experimentally produced by recombination and mutation. Structural and functional aspects of the oncoviruses are given under Retroviridae.

O'Nyong-nyong	alphavirus	Togaviridae
Operophtera brumata (*Lepidoptera*)	B	Entomopoxvirinae
Orangutan herpes (h2)		Gammaherpesvirinae
Orbi[104]: genus of Reoviridae (type species: blue tongue virus)		Reoviridae
Oregon sockeye disease (salmon)	vesiculovirus	Rhabdoviridae

Virus	Group or subgroup	Genus	Subfamily or family
Oreopsyche angustella (Lepidoptera)		B	Entomopoxvirinae
Orf	type species	parapoxvirus	Chordopoxvirinae
Oriboca	C	bunyavirus	Bunyaviridae
Oropouche	Simbu	bunyavirus	Bunyaviridae
Orteca monkey pox		orthopox	Chordopoxvirinae
Orthmonius batesi (Coleoptera)		A	Entomopoxvirinae

ORTHOMYXOVIRIDAE[105]: (Figure 7) This is a monogenic family, the influenza viruses, types A, B, and C. The particles are pleomorphic, about 100 nm in diameter (about 750 S, density in sucrose 1.19 g/cm^3); very large filamentous particles also occur. The lipid-rich envelope carries many of two (or three) types of glycoprotein spikes, the hemagglutinin and the neuraminidase of about 80 and 60×10^3 daltons. The RNA of types A and B is in eight segments of 0.2 to 1×10^6 daltons (total about 5×10^6 daltons); they are minus strands, and carry at least ten genes, the shortest two overlapping ones. They are covered by the nucleocapsid protein as helical filaments of varying lengths. The main protein makes up the matrix that underlies the envelope. Three large proteins are associated with RNA polymerase and other enzymatic functions. These viruses are very sensitive to ether and are inactivated below pH 5. The sequences of all their RNAs and proteins are known.

Influenza viruses, through spontaneous point mutations and reassortment of genome components, particularly of the hemagglutinin, form new strains of different antigenic properties with great frequency. Type A is pathogenic in man, and many of its strains in pigs and horses, others in birds and other animals. Interspecies transfers occur. Transmission is largely by aerosol. Type B appears to occur only in man. Type C may consist of only 7 segments (total less than 5×10^6 daltons). It lacks neuraminidase; the single glycoprotein of 88×10^3 daltons has a receptor-destroying property.

Orthopox[106]: genus of Chordopoxvirinae (type species: vaccinia virus)

Orthoreo[107]: genus of Reoviridae (type species: vesicular stomatitis virus)　　　　　　　Reoviridae

Orungo	human, related to Lebombo	orbivirus	Reoviridae
Oryctes rhinoceros[108]	nonoccluded (C)		Baculoviridae
Orygia pseudotsugata	granulosis and nuclear polyhedrosis viruses (licensed for use as insecticide), (A, B)		Baculoviridae
Ossa	C	bunyavirus	Bunyaviridae
Ostrea edulis (oyster)	unclassified (125–170 nm particles)		
ovi 1–5 (ovis aries, sheep)		mastadenovirus	Adenoviridae
Ovine dependovirus		dependovirus	Parvoviridae
Owl monkey cytomegalo (h4)		cytomegalovirus	Herpesviridae
Oyster	possibly birna virus group		
P (Crustaceae)		orbivirus	Reoviridae
P (Drosophila melanogaster)			Picornaviridae
Pacific pond turtle herpes (h2)			Herpesviridae
Pacui		phlebovirus	Bunyaviridae
Pahayokee	Patois	bunyavirus	Bunyaviridae
Painted turtle herpes (h2)			Herpesviridae
Palestina	Minatitlan	bunyavirus	Bunyaviridae

Virus	Group or subgroup	Genus	Subfamily or family
Palyam (transmitted by *culicoides*)		orbivirus	Reoviridae
Papilloma virus[109] (Figure 10): genus of Papovaviridae (type species: Shope papilloma virus)	type species of group		

PAPOVAVIRIDAE[110]: Two genera make up this family, the larger the PAPILLOMA and the smaller the POLYOMA viruses. Both are icosahedral, of 55 and 45 nm diameter (300 and 240 S), respectively, with a density in CsCl of 1.32 g/cm³). They consist of one molecule of circular double-stranded DNA of 5 and 3×10^6 daltons, and seven and five proteins, respectively. Several polyoma DNAs have been sequenced. Papovaviruses have been found in many mammals, including man (nine types of papilloma and two of polyomaviruses). Their host range for pathogenicity (if any) is very restrictive, but the polyomaviruses are frequently oncogenic in young animals of other species. Papilloma viruses form benign but occasionally also malignant tumors in their host. The type species are Shope rabbit papilloma virus for papillomaviruses, and mouse polyoma virus for polyomavirus, with the simian SV40 being the most studied and used of about ten known members of the genus (Figure 10).

Virus	Group or subgroup	Genus	Subfamily or family
Para: SV40-adenovirus type 7 hybrid			
Parainfluenza 1–4 (human, bovine, ovine, murine)		paramyxovirus	Paramyxoviridae
Parainfluenza 5 (avian, simian, canine)		paramyxovirus	Paramyxoviridae
Parainfluenza turkey Ontario		paramyxovirus	Paramyxoviridae
Paramaecium spp. (protozoon)	unclassified (80 nm particles)		
Paramushir	Sakhalin	probably nairovirus	Bunyaviridae

PARAMYXOVIRIDAE[111]: Pleomorphic virions of about 150–300 nm diameter (about 1000 S, density in sucrose 1.19 g/cm³). A lipid-rich envelope carries 8-nm-long spikes consisting of two glycoproteins (Figure 11). The RNA is a single molecule of about 6×10^6 daltons, mostly minus-stranded. It is built into a flexuous helix that resembles the tobamoviruses (TMV) in its diameter and appearance. There are seven proteins, the largest being the RNA polymerase, the most abundant the matrix supporting the envelope. Of the three genera, the PARAMYXOVIRUSES is the largest with Newcastle disease virus as type species, others being the mumps and several parainfluenza viruses of man, as well as paramyxoviruses infecting other mammals and others birds. Each is very host-specific. All paramyxoviruses have in their spikes a glycoprotein carrying both neuraminidase and hemagglutinating activity. The second spike glycoprotein is active in causing cell fusion. Transmission is usually by aerosol.

The genus MORBILLIVIRUS represents the serologically interrelated measles, the type species, and the canine distemper, rinderpest, and peste-des-petits-ruminants viruses. These lack the neuraminidase.

The type species of the small genus PNEUMOVIRUS is the respiratory syncytial virus and others are bovine respiratory syncytial virus and pneumonia virus of mice. These also lack the neuraminidase, and differ in the size and appearance of their nucleocapsid from the other genera (Figure 11).

Parana	Tacaribe complex	arenavirus	Arenaviridae
Parapox: genus of Chordopoxvirinae (type species Orfvirus)			
Pararota: chicken, porcine, and human rotaviruses with the typical RNA segments, but lacking the common group-specific antigen of other rotaviruses			Reoviridae
Parma wallaby (h1)			Herpesviridae
Paroo River		orbivirus	Reoviridae
Parrot herpes (Pacheco's disease) (h1)			Herpesviridae
Parrot paramyxo		paramyxovirus	Paramyxoviridae
Parry Creek (invertebrates)			probably Rhabdoviridae

PARVOVIRIDAE[112]: (Figure 12) This family consists of small (18–26 nm diameter) icosahedral particles of about 115 S, density in CsCl 1.4 g/cm³, that contain single-stranded DNA of 1.5–2.0 × 10⁶ daltons. The viruses are stable over the range of pH 3–9, and up to 60°C. The three proteins of the virion may result from the processing of a single

Virus	Group or subgroup	Genus	Subfamily or family

translation product, and there are also two nonstructural proteins. The Parvoviridae have three genera: the PARVOVIRUSES of vertebrates, the insect parvoviruses, termed DENSOVIRUSES, and the ADENOASSOCIATED VIRUSES termed DEPENDOVIRUSES. The vertebrate parvoviruses contain usually mostly if not only minus-strand DNA with particular terminal double-hairpin features. They are rarely pathogenic. They replicate only in dividing cells. Several parvoviruses show serological interrelations, i.e., the canine, feline, murine, and racoon viruses. The type species is the Kilham rat virus, or parvovirus rl, and others the minute virus of mice, the Aleutian mink disease virus, the feline panleucopenia, and mink enteritis viruses, H-1, Lu III, RT, and TVX, and possibly the agent causing gastroenteritis of man.

The dependoviruses are, as the name indicates, dependent on coinfection with adenoviruses (or under certain circumstances, herpes virus). These viruses have incapsidated either plus- or minus-stranded DNA of 1.8×10^6 daltons; thus, upon isolation of the DNA, double-stranded molecules result. The viruses are host-specific, but occur in many mammalian adenovirus hosts. They are serologically interrelated.

The densoviruses carry, like the dependoviruses, both plus- and minus-strand DNA. They are named only by the name of their insect host genus. Their virions form dense intranuclear masses, thus expanding the nucleus. The type species is the densovirus of *Junonia*, others of *Aedes, Acheta, Bombyx, Diatrea, Nymphalidae,* and *Sibine.*

Virus	Group or subgroup	Genus	Subfamily or family
Parvo: genus of Parvoviridae (type species: Kilham rat virus)			
Pata (mosquito-transmitted)	Eubenanga	orbivirus	Reoviridae
Patas monkey herpes (h7)			Alphaherpesvirinae
Pathum Thani	Dera Ghazi Khan	nairovirus	Bunyaviridae
Patois	type species of group	bunyavirus	Bunyaviridae
Peaton	Simbu		Bunyaviridae
Penaeus duorarum (shrimp)			Baculoviridae
Percid herpes (h1): *see* Wally epidermal hyperplasia			

Peste-des-petits-ruminants (ovine)	morbillivirus	Paramyxoviridae
Pesti[113]: genus of Togaviridae (type species: bovine viral diarrhea)		
pha 1 (*Phasianus colchicas*, pheasant)	aviadenovirus	Adenoviridae
Phalacrocoracid h1: *see* cormorant herpes virus		
Pheasant leukemia: *see* oncovirus		
Philosamia ricini: nudaurelia β group		
Phlebotomus fever group	unclassified	
Phlebo[114]: genus of Bunyaviridae (type species: sandfly fever Sicilian virus)		
Phnom Penh bat	flavivirus	Togaviridae
Phthorimasa operculetta	cotiavirus	Entomopoxvirinae
Phyllopertha horticola (*Coleoptera*)	A	Entomopoxvirinae
Phytoreoviruses: *see* Section II (plant viruses)		
Pichinde	Tacaribe complex	Arenaviridae

PICORNAVIRIDAE[115]: Small spherical particles of about 27 nm diameter (140–165 S, density in CsCl 1.33–1.44 g/cm^3), consisting of one molecule of plus-strand RNA of 2.5×10^6 daltons with a 5' terminally-bound small protein and 3' terminal poly A, and about equal numbers (60) of four capsid proteins (three of 24–41×10^3, and one of about 10×10^3 daltons). Most of the picornaviruses are very host-specific. The four genera are the ENTEROVIRUSES, CARDIOVIRUSES, RHINOVIRUSES, and APHTHOVIRUSES. The type species of the ENTEROVIRUSES (density 1.34 g/cm^3) is human poliovirus 1; many human coxsackie viruses belong in this genus, as well as murine poliovirus, and enteroviruses of many mammalian species. Human poliovirus RNA has been sequenced (7434 nucleotides). These viruses are stable at acid pH.

They affect mostly the gastrointestinal tract with only occasional neuro- and myotropic tendencies. Enterovirus serotypes 70 and 71 are responsible for human acute hemorrhagic conjunctivitis and encephalitis-meningitis, at times epidemic.

The CARDIOVIRUS type species is the encephalomyocarditis virus and others are the Mengo, Maus–Elberfeld, Columbia SK, MM, and murine encephalomyelitis virus. They differ from the enteroviruses in being unstable below pH 5; their RNA contains a poly(C) tract near, but not at, the 5′ end. They behave identically in sophisticated serologic tests. The RHINOVIRUSES are the main agents of the common cold. Human rhinovirus 1A is the type species, but there exist at least 112 more human, as well as some bovine rhinoviruses. They also are unstable below pH 6. The type member of the APHTHOVIRUSES is the foot-and-mouth disease virus (Aphtho O). Others are A, C, SAT 1, SAT 2, SAT 3, and Asia 1. The aphthoviruses are also unstable below pH 6 and carry a poly(C) tract near the 5′ end. The foot-and-mouth disease virus has a higher buoyant density than the others (1.44 g/cm³). The equine rhinoviruses and several insect viruses are not yet classified picornaviridae, and others are possible members of that family.

Virus	Group or subgroup	Genus	Subfamily or family
Pig cytomegalo (h2)		cytomegalovirus	Betaherpesvirinae
Pig enterovirus		enterovirus	Picornaviridae
Pigeon herpes (h1)			Herpesviridae
Pigeon pox		avipox	Chordopoxvirinae
Pike fry			Rhabdoviridae
Piry (vertebrate hosts)		vesiculovirus	Rhabdoviridae
Pixuna		alphavirus	Togaviridae
Plasmodium gallinaceum (protozoon)	unclassified (35–55 nm particles)		
Playas	Bunyamwera	bunyavirus	Bunyaviridae

Pleuronectid herpes h1: *see* turbot herpes virus

Plodia interpunctella   granulosis (B)   Baculoviridae

Pneumonia of mice   pneumovirus   Paramyxoviridae

Pneumo[116]: genus of Paramyxoviridae (type species: respiratory syncytial virus)

Poikilo[117]: genus of Alphaherpesvirinae (type species: pseudorabies virus)

Poliomyelitis[117] (synonym: polio)   enterovirus   Picornaviridae

Polydnaviridae: proposed family name for insect viruses with segmented double-stranded circular DNA genome.

Polyhedral cytoplasmic DNA virus (polyhedrosisvirus): obsolete term for Iridoviridae

Polyoma[118]: genus of Papovaviridae (type species: SV40 virus)

Pongine herpes h1: *see* herpes pan, chimpanzee herpesvirus

Pongine herpes h2: *see* orangutan herpesvirus

Pongine herpes h3: *see* herpesvirus gorilla

Pongola   Bwamba   bunyavirus   Bunyaviridae

Ponteves   uukuvirus   Bunyaviridae

Poovoot (tick-transmitted)   Kemorovo   orbivirus   Reoviridae

Porcine enteroviruses 1–8   enterovirus   Picornaviridae

Porcine hemagglutinating encephalitis   coronavirus   Coronaviridae

Porcine leukemia: *see* oncovirus

Porcine parvo   parvovirus   Parvoviridae

Virus	Group or subgroup	Genus	Subfamily or family
Porcine transmissible gastroenteritis		coronavirus	Coronaviridae
Powassan (tick-transmitted)		flavivirus	Togaviridae

POXVIRIDAE[119]: This is a family of large complex viruses affecting most vertebrates and many invertebrates. The ovoid particles of about 400 × 200 nm, containing a single molecule of double-stranded DNA of 90–200 × 10⁶ daltons, at least 40 structural proteins, glycoproteins, and phosphoprotein, forming specific internal organelles (lateral bodies, etc.), and lipid that is not in all genera near the surface and thus ether-sensitive. The viruses are generally very host-specific; they are transmitted by aerosol, by contact, and by insects.

The two recognized subfamilies are the CHORDOPOXVIRINAE of vertebrates and the ENTOMOPOXVIRINAE of arthropods. There is some serological relationship between all chordopoxvirinae, and close relationships within each genus. These are: ORTHOPOXVIRUS, formally called the VACCINIA subgroup; PARAPOXVIRUS, the ORF subgroup; AVIPOXVIRUS, the FOWLPOX subgroup; CAPRIPOX virus, formerly sheeppox subgroup; LEPORIPOX VIRUS, the (rabbit) MYXOMA subgroup; and SUIPOXVIRUS, the swinepox virus subgroup. The orthopoxviruses (type species vaccinia virus) are ether-resistant; their linear DNA of 130 × 10⁶ daltons tends to cyclize by terminal crosslinking. Variola of man, ectromelia of mice, and many other mammalian species are known. The parapoxviruses (type species Orf virus) have a somewhat smaller DNA, are ether-sensitive, and show a characteristic surface structure. They infect ungulates, and rarely man. The avipoxviruses (type species fowlpox virus) have a larger DNA, are ether-resistant and generally transmitted by arthropods. The capripox-viruses (type species sheep poxvirus) are transmitted in the same manner and are also ether-sensitive. The leporipox viruses (type species myxoma) show similar properties. The fibromas of rabbits, hares, and squirrels belong to this group. The suipox genus represents the swinepox virus.

The ENTOMOPOXVIRINAE show somewhat different surface appearance, and some have only one lateral body rather than the two of the Chordopoxvirinae. The four probable genera termed A, B, C, and cotiavirus show no serological relationship. Insect poxviruses probably do not replicate in vertebrates, and vice versa. They are named only by their respective hosts. The type species of genusA is *Melothonta melothonta*, with related viruses found in eight other Coleoptera. The type species of genus B is *Amsacta moorei*, with seven other *Lepidoptera* and one *Orthoptera* carrying related viruses. Type species of genus C is *Chironimus luridus*, with related viruses found in five other *Diptera*.

PPR		morbillivirus	Paramyxoviridae
Prague (avian sarcoma): *see* oncovirus			
PRC II (avian sarcoma): *see* oncovirus			
Pretoria	Dera Ghazi Khan	nairovirus	Bunyaviridae
Progressive multifocal leukoencephalopathy agent: *see* JC virus			
Progressive Pneumonia			Lentivirinae
Prospect Hill	serologically related to Hantaan virus		possibly Bunyaviridae
Pseudaletia separata		cotiavirus	Entomopoxvirinae
Pseudaletia unipuncta	granulosis virus (B)		Baculoviridae
Pseudocowpox		orthopoxvirus	Chordopoxvirinae
Pseudolumpy skin disease (bovine)		simplexvirus	Alphaherpesvirinae
Pseudoplusia includens (soybean looper)	nudaurelia β group		
Pseudorabies (swine)		poikilovirus	Alphaherpesvirinae
Psittacid herpes 1: *see* parrot herpesvirus, Pacheco's disease virus			
Pueblo Viejo	Gamboa	bunyavirus	Bunyaviridae
Punta Salinas	Hughes	nairovirus	Bunyaviridae
Punta toro	Sandfly	phlebovirus	Bunyaviridae
Qalyub	type species of group	nairovirus	Bunyaviridae

Virus	Group or subgroup	Genus	Subfamily or family
Quailpox		avipoxvirus	Chordopoxvirinae
Rabbit fibroma		leporipoxvirus	Chordopoxvirinae
Rabbit herpes (h1)			Gammaherpesvirinae
Rabbit kidney vacolating		polyomavirus	Papovaviridae
Rabbit myoxoma		leporipoxvirus	Chordopoxvirinae
Rabbit papilloma: *see* Shope papilloma			
Rabbitpox		orthopoxvirus	Chordopoxvirinae
Rabbit syncytium		probably orbivirus	Reoviridae
Rabies		lyssavirus	Paramyxoviridae
Raccoon pox		probably orthopoxvirus	Chordopoxvirinae
Rana[120]: genus of Iridoviridae (type species: frog viruses)			
Ranid herpes 1: *see* Lucké virus			
Ranid herpes 2: *see* Frog virus 4			
Rat 1: *see* Kilham rat virus			
Rat corona		coronavirus	Coronaviridae
Rat cytomegalo (h2)		muromegalovirus	Betaherpesvirinae
Rauscher murine virus: *see* oncovirus			
RAV (avian leukosis): *see* oncovirus (Rous-associated)			

| Raza | Hughes | nairovirus | Bunyaviridae |
| Razdan | | unclassified | Bunyaviridae |

RD 114: *see oncovirus*

REOVIRIDAE[121]: (Figs. 13, 5) This family is characterized by large icosahedral particles, at times covered by an outer protein shell (60–80 nm diameter, density about 1.38 g/cm³), containing 10–12 molecules of linear doubles-stranded RNA of $0.2–3.0 \times 10^6$ daltons. The internal structure is the nucleocapsid core (45% RNA) with 12 spikes. Each RNA molecule is associated with RNA polymerase, and becomes transcribed in synchrony. There are four animal genera (ORTHOREOVIRUS, ORBIVIRUS, ROTAVIRUS, CYPOVIRUS) and two plant genera (PHYTOREOVIRUS and FIJIVIRUS). The ORTHOREO-VIRUSES occur in many mammals and birds, but are rarely seriously pathogenic. Reovirus type 1 of man is the type species (734 S). They contain ten RNAs of $0.5–2.7 \times 10^6$ daltons, as well as many oligonucleotides. Nine structural proteins of $34–155 \times 10^3$ daltons have been identified. The viruses are quite heat stable and not inactivated by ether or low pH. Besides the human (types 1–3), orthoreoviruses have been found in monkeys, dogs, cattle, and birds; they are serologically related.

The ORBIVIRUSES (550 S) are very numerous in mammals including man and in many insects and other arthropods. They differ from the orthoreoviruses in being sensitive to acid and ether, apparently containing some lipid in their outer shell. The 32 capsomers of the inner shell appear as characteristic rings on the surface of the particle. The orbiviruses are transmitted by and replicated in insects, and many of them are pathogenic only to insects or other invertebrates. The type species is the bluetongue virus of sheep with 21 strains. There are more than 12 other serological subgroups, and several unclassified orbiviruses.

The ROTAVIRUSES (525 S) of mammals have 11 RNA molecules; they are relatively heat-, acid-, and ether-resistant. They are pathogenic in mammals, causing particularly gastrointestinal symptoms. They show little host specificity. They are named only by the host from which they were isolated. The CYPOVIRUSES (type species cytoplasmic poly-hedrosis virus, Figure 5) of insects have again ten genes.

| Reo (man, monkey, etc.) | serotypes 1–3 | orthoreovirus | Reoviridae |

Virus	Group or subgroup	Genus	Subfamily or family
Reptilian sarcoma: see oncovirus			
Respiratory equine		parainfluenza virus	Paramyxoviridae
Respiratory syncytial[122] (widespread child pathogen)		pneumovirus	Paramyxoviridae
Restan	C	bunyavirus	Bunyaviridae
Reticuloendotheliosis: see oncovirus			

RETROVIRIDAE[123]: (Fig. 14) The common feature of this family is that they contain RNA that must be transcribed to DNA, and thus carry a reverse transcriptase in their virion. The virions of about 90 nm diameter, (550–600 S, density in sucrose 1.16–1.21 g/cm^3) are spherical and enveloped by a lipid-rich membrane with spikes containing two glycoproteins (Figure 14). The RNA of the nondefective viruses has a molecular weight of 3.3×10^6, but occurs in the virion as duplex molecules specifically H-bonded to one another (about 70 S). The RNAs are capped, with one methyl group on the chain-terminal nucleotide (7-MeG-5'-p-p-5'MePurine-p→), and have poly(A) on the 3' end. The virion contains tRNAs, with a specific one H-bonded by base complementarity to a specific site on the virion RNA, and serving as primer for the reverse transcriptase. Besides the two envelope proteins and the reverse transcriptase, there are four nucleocapsid and/or matrix proteins, three of which are phosphoproteins, all derived by processing from a primary gene product. Three subfamilies have been distinguished. The largest, the most important, and the best known of which are the ONCOVIRINAE with three genera: ONCOVIRUS B-TYPE, C-TYPE, and D-TYPE based on structural differences. These were formerly termed RNA tumor viruses and will be discussed under oncoviruses in more detail. They have long been recognized to be oncogenic in birds and mammals, recently also at times in man. This property could be detected in cell culture by transformation of cells leading to microtumors. Three subgenera are the mammalian, avian, and reptilian C-type oncoviruses.

The SPUMAVIRINAE or FOAMY VIRUS GROUP, while having similar appearance and replication properties as the oncovirinae, do not transform cells nor cause tumors, nor are they usually pathogenic. They lead to persistent infections in several mammals, characterized by the appearance under specific conditions of vacuolated and, thus, seemingly foamy cell syncytia. They are thus termed syncytial or foamy viruses.

The LENTIVIRINAE resemble the other subfamilies in all biochemical respects. However, they cause slowly progressive lethal diseases in sheep, etc., and possibly man. They are not serologically related to other oncovirinae.

RFV (variant of BK) (contains two complementary DNA molecules), (defective human)	polyomavirus	Papovaviridae

RHABDOVIRIDAE[124]: This family includes both animal and plant viruses. The animal rhabdoviruses are generally bullet-shaped while many plant rhabdoviruses are rounded at both ends, thus bacilliform. The virions vary in dimension, from 130–380 × 50–95 nm (550–1000 S, density in CsCl or sucrose 1.19 g/cm^3). They are unstable below pH 4, at 56°C, and in ether. The nucleocapsid contains negative-strand RNA of about 4 × 10^6 daltons, in a protein (70% of total) helix, and the RNA polymerase, possibly consisting of both a large and a medium-sized protein. The lipid-rich envelope contains a glycoprotein.

The animal rhabdoviridae generally have wide host ranges, many of them replicating in insects and vertebrates, some even in insects and plants. They are mostly transmitted by insects. The vesicular stomatitis virus is the type species of the genus VESICULOVIRUSES; there are a few other members, besides several strains of that virus. Many rhabdoviruses have not yet been classified in terms of genera. The rabies virus and some similar ones have, however, been classified as a separate genus, LYSSAVIRUSES, on serological grounds. These again were found partly in vertebrates and partly in insects, but are not restricted to one host species (Figure 15).

Rhadino: genus of Gammaherpesvirinae		Gammaherpesvirinae
Rhesus leucocyte-associated (Strain I) (h11)		Gammaherpesvirinae
Rhesus leucocyte-associated (Strain II) (h12)		Betaherpesvirinae
Rhesus monkey cytomegalo (h8)		Betaherpesvirinae
Rhino[125]: genus of Picornaviridae (type species: common cold virus)		
Rhinopneumonitis: *see* equine abortion virus		
Rift Valley fever	phlebovirus	Bunyaviridae

Virus	Group or subgroup	Genus	Subfamily or family
Rinderpest		morbillivirus	Paramyxoviridae
Rio Bravo		flavivirus	Togaviridae
Rio Grande		phlebovirus	Bunyaviridae
RKV (rabbit)		papillomavirus	Papovaviridae
RNA tumor virus group: *see* Oncovirinae			
Rochester-2-sarcoma (avian): *see* oncovirus			
Rocio		flavivirus	Togaviridae
Roseola infantum agent	unclassified		
Ross River		alphavirus	Togaviridae
Rota[126]: genus of Reoviridae (agent of human epidemic polyarthritis), 11 segment of dsRNA, many hosts			
Rous-associated viruses (RAV): *see* oncovirus			
Rous sarcoma (avian)[127]: *see* oncovirus			
Royal Farm (tick-borne)		flavivirus	Togaviridae
RPL 12 (avian leukosis): *see* oncovirus			
RT (rat)		parvovirus	Parvoviridae
Rubella[128] (German measles)	type species	rubivirus	Togaviridae
Rubi: genus of Togaviridae (type species: rubella virus)			
Russian tick-borne encephalitis		flavivirus	Togaviridae

RV: *see* Kilham rat viruses			
S 1643 (invertebrate hosts)			Rhabdoviridae
SA 6 (monkey) (h3)			Betaherpesvirinae
SA 7 (monkey) (oncogenic simian adenovirus)		mastadenovirus	Adenoviridae
SA 8 (monkey) (h2)		simplexvirus	Alphaherpesvirinae
SA 11 (monkey)		rotavirus	Reoviridae
SA 12 (baboon)		polyomavirus	Papovaviridae
SA 15 (monkey) (h4)			Betaherpesvirinae
Sabo	Simbu	bunyavirus	Bunyaviridae
Saboya		flavivirus	Togaviridae
Sac brood		unclassified, similar to rhinovirus, but much lower buoyant density (1.33 g/cm³)	
Sacramento River chinook			probably Rhabdoviridae
Sagiyama		alphavirus	Togaviridae
Saimiriine herpes 1: *see* Marmoset herpes virus, herpes M, T, tamavirus, platyrrhinae type 1			
Saimiriine herpes 2: *see* squirrel monkey, herpes saimiri.			
Saint Floris		probably phlebovirus	Bunyaviridae
Saint Louis encephalitis: *see* St. Louis			

Virus	Group or subgroup	Genus	Subfamily or family
Sakhalin	type species of group	nairovirus	Bunyaviridae
Salehabad		phlebovirus	Bunyaviridae
Salivary of bats		flavivirus	Togaviridae
Salmon herpes h1: *see* salmonid herpes			
Salmon herpes h2: *see Oncorhynchus masou*			
Salmonid herpes (h1)			Herpesviridae
San Angelo	California	bunyavirus	Bunyaviridae
Sandfly fever Sicilian	type species of group	phlebovirus	Bunyaviridae
Sango	Simbu	bunyavirus	Bunyaviridae
San Juan	probably Gamboa	bunyavirus	Bunyaviridae
San Miguel sea lion	(at least eight serotypes)		Caliciviridae
Santa Rosa	Bunyamwera	bunyavirus	Bunyaviridae
Sapphire 11	Hughes	nairovirus	Bunyaviridae
Sarracenia purpurea			probably Rhabdoviridae
SAT 1, SAT 2, SAT 3		aphthoviruses	Picornaviridae
Satellite: a term used in plant virology for "helper"-dependent viruses			
Sathuperi	Simbu	bunyavirus	Bunyaviridae

Name	Genus	Family
Saumarez Reef (tick-transmitted)	flavivirus	Togaviridae
Sawgrass (invertebrate hosts)		probably Rhabdoviridae
Schmidt–Ruppin avian sarcoma: *see* oncovirus		
Scrapie agent[129]	unclassified (possibly nonviral)	proposed term: Prion
Scurid herpes h1: *see* European ground squirrel cytomegalovirus		
SD	murine coronavirus strain (*see* SK)	Coronaviridae
Seletar (tick-transmitted)	Kemerovo orbivirus	Reoviridae
Semliki Forest	alphavirus	Togaviridae
Sendai (synonym for parainfluenza virus 1)	paramyxovirus	Paramyxoviridae
Sepia officinalis (squid)	paramyxovirus	Paramyxoviridae
Sepik (mosquito-borne)	flavivirus	Togaviridae
Sericestis iridescent	iridovirus	Iridoviridae
Serra do Navio	California bunyavirus	Bunyaviridae
SF Naples	phlebovirus	Bunyaviridae
SF Sicilian: *see* sandfly virus		
Shamonda	Simbu bunyavirus	Bunyaviridae
Shark River	Patois bunyavirus	Bunyaviridae

Virus	Group or subgroup	Genus	Subfamily or family
Sheep herpes (h1)	type species		Herpesviridae
Sheeppox		capripoxvirus	Chordopoxvirinae
Sheep pulmonary adenomatosis			Herpesviridae
Shipping fever: *see* parainfluenza virus			
Shope fibroma (rabbit)		leporipoxvirus	Chordopoxvirinae
Shope papilloma (rabbit)		papillomavirus	Papovaviridae
Shuni	Simbu	bunyavirus	Bunyaviridae
Sialodacryoadenitis (rat)		coronavirus	Coronaviridae
Siamese cobra herpes			Herpesviridae
Si Ar 126	possibly Thogato	bunyavirus	Bunyaviridae
Sibine		densovirus	Parvoviridae
Sigma (*Drosophila*)			probably Rhabdoviridae
Silverwater	Kaisodi	bunyavirus	Bunyaviridae
Simbu	type species of group	bunyavirus	Bunyaviridae
Simian (monkey) adeno (oncogenic)	SV (1, 22, 23, 25, 33, 37, 38)	adenoviruses	Adenoviridae
Simian B: *see* simian herpes B virus			
Simian entero		enterovirus	Picornaviridae

Name	Description	Genus	Family
Simian foamy			Spumavirinae
Simian hemorrhagic fever		possibly pestivirus	Togaviridae
Simian herpes B (h1)			Alphaherpesvirinae
Simian paramyxo	SV 5, SV 41	paramyxovirus	Paramyxoviridae
Simian polyoma (SV 40, sequenced and much used)		polyomavirus	Papovaviridae
Simian reo	SV 12, SV 59, identical with human type 1, and 2 reovirus	orthoreovirus	Reoviridae
Simian sarcoma: *see* oncovirus			
Simian varicella: *see* Medical Lake macaque			
Simplex: genus of Alphaherpesvirinae (type species: human herpes simplex 1)			
Sindbis[130] (Figure 16)	type species	alphavirus	Togaviridae
Sixgun City (tick-borne)	Kemorovo	orbivirus	Reoviridae
SK (mouse)	(strain derived from human multiple sclerosis CNS tissue)	coronavirus	Coronaviridae
SL3-3 (mouse leukemia): *see* oncovirus			
Slow-growing cytomegalo (of horses) (h2)			Betaherpesvirinae
Slow paralysis: *see* bee slow paralysis virus			
Slow viruses: term used for Lentivirinae, and at times (falsely) for the scrapie, Kuru, Creutzfeldt–Jacob, transmissible mink encephalopathy agent group (proposed term: Prions).			

Virus	Group or subgroup	Genus	Subfamily or family
SM feline sarcoma: *see* oncovirus			
Snowshoe hare	California	bunyavirus	Bunyaviridae
Sokoluk		flavivirus	Togaviridae
Soldado	Hughes	nairovirus	Bunyaviridae
Sororoca	Bunyamwera	bunyavirus	Bunyaviridae
South river	probably California	bunyavirus	Bunyaviridae
Soybean looper	unclassified 40 nm particle (190S) with 1.9×10^6 dalton RNA and 55×10^3 dalton protein, serologically unrelated to other insect viruses tested		
Sp Ar 2317	Anopheles A	bunyavirus	Bunyaviridae
Sparrowpox		avipoxvirus	Chordopoxvirinae
Spider monkey herpes (h1)			Alphaherpesvirinae
Spleen focus-inducing (mink, murine)[131]: *see* oncovirus			
Spleen necrosis[132] (duck)		oncovirus	Retroviridae
Spodoptera exempta	types 3 and 12	cypovirus	Reoviridae
Spodoptera exigua	type species, type 11	cypovirus	Reoviridae

Virus	Group or subgroup	Genus	Subfamily or family
Suipox[135]: genus of Chordopox virinae (type species: swinepox)			
Sunday Canyon		bunyavirus	Bunyaviridae
sus 1–4 (*Sus domesticus*, pig)		mastadenovirus	Adenoviridae
SV (1, 5, 12, 15, 20, 23, 25, 33, 34, 37, 38, 40, 41, 59): *see* under simian adeno, paramyxo, polyoma, reoviruses			
SV 40[136]: *see* simian polyoma virus			
Swim bladder inflammation agent (carp)			probably Rhabdoviridae
Swine fever		pestivirus	Togaviridae
Swine influenza		type A	Orthomyxoviridae
Swinepox	type species	suipoxvirus	Chordopoxvirinae
Swine vesicular disease	(closely related to human coxsackie virus B5)	enterovirus	Picornaviridae
Syncytium-forming: *see* under bovine, feline, etc., syncytium-forming viruses			
T50616 (Skunk)		orbivirus	Reoviridae
T6–20		ranavirus	Iridoviridae
T21 (*Xenopus*)		ranavirus	Iridoviridae
Tacaiuma	Anopheles A	bunyavirus	Bunyaviridae
Tacaribe	type species of group	arenavirus	Arenaviridae

Tadpole edema (*Rana catesbriana*)	LT (1–4, 6–20)	ranavirus	Iridoviridae
Taggert	Sakhalin	nairovirus	Bunyaviridae
Tahyna	California	bunyavirus	Bunyaviridae
Taiassui	Bunyamwera	bunyavirus	Bunyaviridae
Talfan's disease		enterovirus	Picornaviridae
Tamdy			Bunyaviridae
Tamiami		arenavirus	Arenaviridae
Tanapox (monkey)	serologically related to Yaba pox virus	orthopoxvirus	Chordopoxvirinae
Tataguine			Bunyaviridae
Tehran		phlebovirus	Bunyaviridae
Tellina tennis (mollusc)	probably birnavirus group		
Tembusu (mosquito-borne)		flavivirus	Togaviridae
Tensaw	Bunyamwera	bunyavirus	Bunyaviridae
Termite paralysis			Picornaviridae
Teschen's disease: *see* Talfan's disease			
Tete	type member of group	bunyavirus	Bunyaviridae
Theiler feline sarcoma: *see* oncovirus			
Theiler's murine encephalitis: *see* murine encephalomyelitis			

Virus	Group or subgroup	Genus	Subfamily or family
Thetalymphocryto: genus of gammaherpesvirinae (type species: Marek's disease virus)			
Thimiri		bunyavirus	Bunyaviridae
Thirlmere Reservoir	probably birnavirus group		
Thogoto (tick-borne)			possibly Orthomyxoviridae
Thosea asigna	nudaurelia β group		
Tick-borne encephalitis		flavivirus	Togaviridae
Tillamook	Sakhalin	nairovirus	Bunyaviridae
Tilligerry (NB 7080)	Eubenangee	orbivirus	Reoviridae
Timbo (vertebrate-hosts)			probably Rhabdoviridae
Timboteua	Guama	bunyavirus	Bunyaviridae
Tinaroo	Simbu	bunyavirus	Bunyaviridae
Tindholmur (tick-borne)	Kemorovo	orbivirus	Reoviridae
Tipula iridescent	type species	iridovirus	Iridoviridae
Tipula paludosa		iridovirus	Iridoviridae
Tlacotalpan	Bunyamwera	bunyavirus	Bunyaviridae

TOGAVIRIDAE[137]: (Figure 16) These are the smallest enveloped (toga = mantle) animal viruses. They are spherical, 40–70 nm in diameter, density is CsCl 1.25 g/cm^3. The icosahedral nucleocapsid of about 30 nm diameter consists of a 4.0–

4.5×10^6 dalton plus-strand RNA and the capsid protein, the envelope of two glycoproteins, and lipid. Four genera have been identified, members of which are serologically interrelated, but not with members of other genera. These have been called ALPHAVIRUSES, FLAVIVIRUSES, RUBIVIRUSES, and PESTIVIRUSES. The type member of the ALPHAVIRUSES is the Sindbis virus (70 nm diameter, 280 S), the capsid protein of which is about 32×10^3, and the glycoproteins 52×10^3 daltons; the RNA 4.5×10^6 daltons, and 30% lipid. Among the many other members are the Semliki Forest virus, and Eastern, Western, and Venezuelan equine encephalomyelitis viruses. These viruses replicate in both arthropods and vertebrates.

The FLAVIVIRUSES are smaller (45 nm diameter, 200 S) with one glycoprotein of about 60, and a coat protein of 14 $\times 10^3$ daltons. Most of these viruses replicate in both insects and vertebrates. The yellow fever virus is the type species (RNA of 3.8×10^6 daltons, lacking terminal poly A), and Dengue and Japanese encephalitis are other examples of this numerous group. Their replication strategy differs from that of the alphaviruses, and they may become classified as a separate family.

The only classified member of the RUBIVIRUSES is the rubella virus of man, which has no other vertebrate or invertebrate hosts. The PESTIVIRUS genus also differs from the other genera in being unable to replicate in invertebrates, and in lack of serological relation to other togaviridae. The type species is the mucosal disease virus (bovine virus diarrhaea). Hog cholera and border disease viruses are members. Among possible togaviridae (equine arteritis, lactic dehydrogenase, simian hemorrhagic fever viruses, and *aedes albopictus* cell-fusing agent) is also one plant virus, the carrot mottlevirus.

Tonate	alphavirus	Togaviridae
Toluca 1	enterovirus	Picornaviridae
TO (Theiler's Original): *see* murine encephalomyelitis virus		
Toscana	phlebovirus	Bunyaviridae
Transmissible enteritis of turkey	enterovirus	Picornaviridae
Transmissible gastroenteritis of pigs	coronavirus	Coronaviridae

Virus	Group or subgroup	Genus	Subfamily or family
Transmissible mink encephalopathy agent	unclassified (similar to scrapie)		proposed term: Prion
Treeshrew herpes			Herpesviridae
Tribec (tick-borne)	Kemorovo	orbivirus	Reoviridae
Trichoplusia ni	granulosis virus (B)		Baculoviridae
Trichoplusia ni	nudaurelia β group		
Trichoplusia ni	type 5	cypovirus	Reoviridae
Tricosomoides crasscanda (nematode)	unclassified (15 nm particles)		
Triphena pronuba	type 7	cypovirus	Reoviridae
Trivittatus	California	bunyavirus	Bunyaviridae
Trubanaman	Maputta	bunyavirus	Bunyaviridae
Tupaiine herpes h1: *see* Treeshrew herpes			
Turbot herpes	pleuronectid	herpesvirus	Herpesviridae
Turkeypox (h1) (used to vaccinate against Marek's disease)		thetalymphocrytovirus	Gammaherpesvirinae
Turlock	type member of group	bunyavirus	Bunyaviridae
TVS		parvovirus	Parvoviridae

Type B, C, D: B-type, C-type, D-type: subgenera of oncovirus

Tyuleniy (tick-borne)		flavivirus	Togaviridae
Uganda S (mosquito-borne)		flavivirus	Togaviridae
UK bovine rota (related to SA 11)		rotavirus	Reoviridae
Umatilla		orbivirus	Reoviridae
Umbre	Turlock	bunyavirus	Bunyaviridae
Una		alphavirus	Togaviridae
Urucuri		phlebovirus	Bunyaviridae
USA T5-0616		orbivirus	Reoviridae
USA 69-V2161		orbivirus	Reoviridae
US bat salivary		flavivirus	Togaviridae
Usutu (mosquito-borne)		flavivirus	Togaviridae
Utinga	Simbu	bunyavirus	Bunyaviridae
Utive	Simbu	bunyavirus	Bunyaviridae

Uuku[138]: genus of Bunyaviridae (type species: uukuniemivirus)

Uukuniemi	type species	uukuvirus	Bunyaviridae
Vaccinia[139] (Figure 17)		orthopoxviruses	Chordopoxvirinae

Varicella[140]: genus of alphaherpesvirinae (type species: varicella-zoster)

Varicella–zoster (chickenpox, shingles)		varicellavirus	Alphaherpesvirinae

Virus	Group or subgroup	Genus	Subfamily or family
Variola (minor) (human)		orthopoxvirus	Chordopoxvirinae
Vellore	Palyam	orbivirus	Reoviridae
Venezuelan (equine) encephalitis		alphavirus	Togaviridae
Vesicular exanthema of swine	type species (12 serotypes)		Caliciviridae
Vesicular stomatitis	type species (strains: Argentina, Brazil, Cocal, Indiana, New Jersey, etc.) (Figure 15)	vesiculovirus	Rhabdoviridae
Vesiculo[141]: genus of Rhabdoviridae (type species: vesicular stomatitis virus)			
Vinces	probably C	bunyavirus	Bunyaviridae
Viper (reptilian sarcoma): see oncovirus			
Viral hemorrhagic septicemia: see Egtvedvirus			
Virgin River	Anopheles A	bunyavirus	Bunyaviridae
Visna[142] (so-called slow virus)			Lentivirinae
von Magnus (defective influenza)		type B	Orthomyxoviridae
Wad Wedani	Kemorovo	orbivirus	Reoviridae
Wallal (Ch 12048)	type species of subgroup (2 serotypes)	orbivirus	Reoviridae

Name		Genus	Family
Wally epidermal hyperplasia (fish) (h1)	percid		Herpesviridae
Warrego (Ch 9935) (*Culicoides*)	type member of subgroup (2 serotypes)	orbivirus	Reoviridae
Wart (human)		papillomavirus	Papovaviridae
Wesselsbron (mosquito-borne)		flavivirus	Togaviridae
Western equine encephalitis (WEE)		alphavirus	Togaviridae
WF-1 (rat sarcoma): *see* oncovirus			
Whataroa		alphavirus	Togaviridae
Wildebeest malignant catarrhal fever (h1)			Gammaherpesvirinae
Wiseana cervinata		cotiavirus	Entomopoxvirinae
Witwatersrand		bunyavirus	Bunyaviridae
Wongal	Koongol	bunyavirus	Bunyaviridae
Woodchuck hepatitis[143]: very similar to human hepatitis B virus			
Woolley monkey sarcoma: *see* oncovirus			
Wyeomyia	Bunyamwera	bunyavirus	Bunyaviridae
X 14		parvovirus	Parvoviridae
XBM (bovine)		orbivirus	Reoviridae
Xenopis T21		ranavirus	Iridoviridae
Y 73 (avian leukosis): *see* oncovirus			

Virus	Group or subgroup	Genus	Subfamily or family
Yaba 1	Turlock	bunyavirus	Bunyaviridae
Yaba 7	Simbu	bunyavirus	Bunyaviridae
Yaba monkey tumor pox	related to tanavirus	orthopoxvirus	Chordopoxvirinae
Yamaguchi sarcoma (avian): *see* oncovirus			
Yaquine Head (tick-borne)	Kemorovo	orbivirus	Reoviridae
Yata (invertebrate host)			probably Rhabdoviridae
Yellow Fever	type species	flavivirus	Togaviridae
Yucaipa (avian)		paramyxovirus	Paramyxoviridae
Zaliv–Terpeniya		uukuvirus	Bunyaviridae
Zegla	Patois	bunyavirus	Bunyaviridae
Zika (mosquito-borne)		flavivirus	Togaviridae
Zirqa	Hughes	nairovirus	Bunyaviridae
Zwogerziekte			Lentivirinae
6/94	parainfluenza virus 1	paramyxovirus	Paramyxoviridae
73U11	C	bunyavirus	Bunyaviridae
75V-2621, 78V-2441, 75V-2374	Gamboa	bunyavirus	Bunyaviridae

References (Section I)

1. Goff, P., Gilboa, E., Witte, O. N., and Baltimore, I., 1980, Structure of the Abelson murine leukemia virus genome and the homologous cellular gene: Studies with cloned viral DNA, *Cell* **22:**727.
2. Grose, C., and Horwitz, M. S., 1976, Characterization of an enterovirus associated with acute infectious lymphocytosis. *J. Gen. Virol.* **30:**347–355.
3a. Berns, K. I., and Hauswirth, W. W., 1979, Adeno-associated viruses, *Adv. Virus Res.* **25:**407.
3b. Matthews, R. E. F., 1982, Classification and nomenclature of viruses, *Intervirology* **17:**73.
4a. Philipson, L., and Lindberg, U., 1974, Reproduction of adenoviruses, in: *Comprehensive Virology*, Vol. 2 (H. Fraenkel-Conrat and R. R. Wagner, eds.), p. 143, Plenum Press, New York.
4b. Ginsberg, H. S., 1979, Adenovirus structural proteins, in: *Comprehensive Virology*, Vol. 13 (H. Fraenkel-Conrat and R. R. Wagner, eds.), p. 409, Plenum Press, New York.
4c. Ginsberg, H. S., and Young, C. S. H., 1979, Genetics of adenoviruses, in: *Comprehensive Virology*, Vol. 9 (H. Fraenkel-Conrat and R. R. Wagner, eds.), p. 27, Plenum Press, New York.
4d. Patch, C. T., Levine, A. S., and Lewis, A. M., Jr., 1979, The adenovirus-SV40 hybrid viruses, in: *Comprehensive Virology*, Vol. 13 (H. Fraenkel-Conrat and R. R. Wagner, eds.), p. 459, Plenum Press, New York.
4e. Matthews, R. E. F., 1982, Classification and nomenclature of viruses, *Intervirology* **17:**59.
4f. Wigand *et al.*, 1982, Adenoviridae: Second report, *Intervirology* **18:**169.
5. Jurkovičová, M., Van Touw, J. H., Sussenbach, J. S., and Ter Schegget, J., 1979, Characterization of the nuclear polyhedrosis virus DNA of *Adoxophyes orana* and of *Barathra brassicae*, *Virology* **93:**8.
6a. Salas, M. L., Kuznar, J., and Vinuela, E., 1981, Polyadenylation, methylation, and capping of the RNA synthesized *in vitro* by African swine fever virus, *Virology* **113:**484.
6b. Goorha, R., and Granoff, A., 1979, Icosahedral cytoplasmic deoxyriboviruses, in: *Comprehensive Virology*, Vol. 14 (H. Fraenkel-Conrat and R. R. Wagner, eds.), p. 347, Plenum Press, New York.
7a. Shahrabadi, M. S., Cho, H. J., and Marusyk, R. G., 1977, Characterization of the protein and nucleic acid of Aleutian disease virus, *J. Virol.* **23:**353.
7b. Porter, D. D., and Cho, H. J., 1980, Aleutian disease of mink: A model for persistent infection, in: *Comprehensive Virology*, Vol. 16 (H. Fraenkel-Conrat and R. R. Wagner eds.), p. 233, Plenum Press, New York.
8. Matthews, R. E. F., 1982, Classification and nomenclature of viruses, *Intervirology* **17:**97.
9. Matthews, R. E. F., 1982, Classification and nomenclature of viruses, *Intervirology* **17:**48.

10. D'Arcy, C. J., Burnett, P. A., Hewings, A. D., and Goodman, R. M., 1981, Purification and characterization of a virus from the aphid *Rhopalosiphum padi*, *Virology* **112**:346.

11. Matthews, R. E. F., 1982, Classification and nomenclature of viruses, *Intervirology* **17**:130.

12a. Rawls, W. E., and Leung, W-C., 1979, Arenaviruses, in: *Comprehensive Virology*, Vol. 14 (H. Fraenkel-Conrat and R. R. Wagner eds.), p. 157, Plenum Press, New York.

12b. Matthews, R. E. F., 1982, Classification and nomenclature of viruses, *Intervirology* **17**:119.

13. Bailey, L., and Woods, R. D., 1974, Three previously undescribed viruses for the honey bee, *J. Gen. Virol.* **25**:175.

14. Herring, A. J., Gray, E. W., and Snodgrass, D. R., 1981, Identification and characterization of ovine astrovirus, *J. Gen. Virol.* **53**:47.

15. Miller, L. K., and Dawes, K. P., 1979, Physical map of the DNA genome of *Autographa californica* nuclear polyhedrosis virus, *J. Virol.* **29**(March):1044.

16. Davies, H. A., Dourmashkin, R. R., and MacNaughton, M. R., 1981, Ribonucleoprotein of avian infectious bronchitis virus, *J. Gen. Virol.* **53**:67.

17a. Czernilofsky, A. P., Levinson, A. D., Varmus, H. E., and Bishop, J. M., 1980, Nucleotide sequence of an avian sarcoma virus oncogene (*src*) and proposed amino acid sequence for gene product, *Nature* **287**.

17b. Erikson, E., Collett, M. S., and Erikson, R. L., 1978, *In vitro* synthesis of a functional avian sarcoma virus transforming-gene product, *Nature* **274**(August).

18. Matthews, R. E. F., 1982, Classification and nomenclature of viruses, *Intervirology* **17**:43.

19a. Matthews, R. E. F., 1982, Classification and nomenclature of viruses, *Intervirology* **17**:53.

19b. Tinsley, T. W., and Harrap, K. A., 1978, Viruses of invertebrates, in: *Comprehensive Virology*, Vol. 12 (H. Fraenkel-Conrat and R. R. Wagner, eds.), p. 1, Plenum Press, New York.

19c. Kelly, D. C., Lescott, T., Ayres, M. E., Carey, D., Coutts, A., and Harrap, K. A., 1981, Induction of a nonoccluded baculovirus persistently infecting *Heliothis zea* by *Heliothis armigera* and *Trichoplusia ni* nuclear polyhedrosis viruses, *Virology* **112**:174.

19d. Langridge, W. H. R., 1981, Biochemical properties of a persistent nonoccluded baculovirus isolated from *Heliothis zea* cells, *Virology* **112**:770.

19e. Revet, B. M. J., and Guelpa, B., The genome of a baculovirus infecting *Tipula paludosa* (Meig) (diptera): A high molecular weight closed circular DNA of zero superhelix density.

19f. McCarthy, W. J., Mercer, W. E., and Murphy, T. F., 1978, Characterization of the DNA from four heliothis nuclear polyhedrosis virus isolates, *Virology* **90**:374.

20a. Bailey, L., 1976, Viruses attacking the honey bee, *Adv. Virus Res.* **20**:271.

20b. Bailey, L., Carpenter, J. M., and Woods, R. D., 1981, Properties of a filamentous virus of the honey bee (*Apis mellifera*), *Virology* **114**:1.

20c. Bailey, L., Carpenter, J. M., Govier, D. A., and Woods, R. D., 1980, Bee virus X, *J. Gen. Virol.* **51**:405.

20d. Longworth, J. F., 1978, Small isometric viruses of invertebrates, *Adv. Virus Res.* **23**:103.

21. McPhee, D. A., and Westaway, E. G., 1981, Comparisons of belmont virus, a possible bunyavirus unique to Australia, with bunyamwera virus, *J. Gen. Virol.* **54**:135.

22. Weiss, M., Steck, F., and Horzinek, M. C., 1983, Purification and partial characterization of a new enveloped RNA virus (berne virus), *J. Gen. Virol.* **64**:1849.

23. Matthews, R. E. F., 1982, Classification and nomenclature of viruses, *Intervirology* **17**:49.

24. MacDonald, R. D., and Gower, D. A., 1981, Geno and phenotypic divergence among three serotypes of aquatic birnaviruses (infectious pancreatic necrosis virus), *Virology* **114**:187.

25. Wright, P. J., and Di Mayorca, G., 1975, Virion polypeptide composition of the human papovavirus BK: Comparison with simian virus 40 and polyoma virus, *J. Virol.* **15:**828.

26. Friesen, P. D., and Rueckert, R. R., 1982, Black beetle virus: Messenger for protein B is a subgenomic viral RNA, *J. Virol.* **42:**986.

27. Della-Porta, A. J., and Brown, F., 1979, Physicochemical characterization of bovine ephemeral fever virus as a member of the family *Rhabdoviridae, J. Gen. Virol.* **44:**99.

28. Ludwig, H., 1983, Bovine herpesviruses in: *The Viruses,* Vol. II (H. Fraenkel-Conrat, R. R. Wagner, and B. Roizman eds.), p. 135, Plenum Press, New York.

29. Buchman, T. G., and Roizman, B., 1978, Anatomy of bovine mammillitis DNA: Restriction endonuclease maps of four populations of molecules that differ in the relative orientation of their long and short components, *J. Virol.* **25:**395.

30. Pritchett, R., Manning, J. S., and Zee, Y. C., and Characterization of bovine viral diarrhea virus RNA, *J. Virol.* **15:**1342.

31a. Bishop, D. H. L., and Shope, R. E., 1979, Bunyaviridae, in: *Comprehensive Virology,* Vol. 14 (H. Fraenkel-Conrat and R. R. Wagner, eds.), p. 1, Plenum Press, New York.

31b. Matthews, R. E. F., 1982, Classification and nomenclature of viruses, *Intervirology* **17:**115.

31c. Beaty, B. J., Holterman, M., Tabachnick, W., and Shope, R. E., 1981, Molecular basis of bunyavirus transmission by mosquitoes: Role of the middle-sized RNA segment, *Science* **211:**1433.

31d. Beaty, B. J., Miller, B. R., Shope, R. E., Rozhon, E. J., and Bishop, D. H. L., 1982, Molecular basis of bunyavirus *per os* infection of mosquitoes: Role of the middle-sized RNA segment, *Proc. Natl. Acad. Sci. USA* **79:**1295.

31e. Obijeski, J. F., and Murphy, F. A., 1977, Bunyaviridae: Recent biochemical developments, *J. Gen. Virol.* **37:**1.

32. Bridger, J. C., and Woode, G. N., 1976, Characterization of two particle types of calf rotavirus, *J. Gen. Virol.* **31:**245.

33a. Schaffer, F. L., 1979, Caliciviruses, in: *Comprehensive Virology,* Vol. 14 (H. Fraenkel-Conrat and R. R. Wagner, eds.), p. 249, Plenum Press, New York.

33b. Schaffer, F. L. *et al.,* 1980, Caliciviridae, *Intervirology* **14:**1.

33c. Matthews, R. E. F., 1982, Classification and nomenclature of viruses, *Intervirology* **17:**133.

34a. Goldman, N., Presser, I., and Sreevalsan, T., 1977, California encephalitis virus: Some biological and biochemical properties, *Virology* **76:**352.

34b. Vorndam, A. V., and Trent, D. W., 1979, Oligosaccharides of California encephalitis viruses, *Virology* **95:**1.

35. Parrish, C. R., and Carmichael, L. E., 1983, Antigenic structure and variation of canine parvovirus type-2, feline panleukopenia virus, and mink enteritis virus, *Virology* **129:**401.

36. Roberson, N. M., McGuire, T. C., Klevjer-Anderson, P., Gorham, J. R., and Cheevers, W. P., 1982, Caprine arthritis-encephalitis virus is distinct from visna and progressive pneumonia viruses as measured by genome sequence homology, *J. Virol.* **44:**755.

37. Matthews, R. E. F., 1982, The capripoxviruses, *Intervirology* **17:**44.

38. Matthews, R. E. F., 1982, The cardioviruses, *Intervirology* **17:**130.

39a. Dixon, R. A. F., and Farber, F. E., 1980, Channel catfish virus: Physicochemical properties of the viral genome and identification of viral polypeptides, *Virology* **103:**267–278.

39b. McAllister, P. E., 1979, Fish viruses and viral infections, in: *Comprehensive Virology,* Vol. 14 (H. Fraenkel-Conrat and R. R. Wagner, eds.), p. 401, Plenum Press, New York.

40. Matthews, R. E. F., 1982, The chloriridoviruses, *Intervirology* **17:**57.

41. Matthews, R. E. F., 1982, The chordopoxvirinae, *Intervirology* **17:**42.

42a. Matthews, R. E. F., 1982, The Coronaviridae, *Intervirology* **17:**102.

42b. Robb, J. A., and Bond, C. W., 1979, Coronaviridae, in: *Comprehensive Virology,* Vol. 14 (H. Fraenkel-Conrat and R. R. Wagner, eds.), p. 193, Plenum Press, New York.

42c. Tyrell, D. A. J., Alexander, D. J., Almeida, J. D., Cunningham, C. H., Easterday, B. C., Garwes, D. J., Hierholzer, J. C., Kapikian, A., MacNaughton, M. R., and McIntosh, K., 1978, Coronaviridae: Second report, *Intervirology* **10:**321–328.

42d. Mahy, B. W. J., 1980, Coronavirus comes of age, *Nature* **288:**536.

42e. Dennis, D. E., and Brian, D. A., 1982, RNA-dependent RNA polymerase activity in coronavirus-infected cells, **42:**153–164.

42f. Siddell, S. G., *et al.* 1983, Coronaviridae, *Intervirology* **20:**181.

43. Ueda, Y., Dumbell, K. R., Tsuruhara, T., and Tagaya, I., 1978, Studies on cotia—an unclassified poxvirus, *J. Gen. Virol.* **40:**263–276.

44a. Wolinsky, J. S., and Johnson, R. T., 1980, Role of viruses in chronic neurological diseases, in: *Comprehensive Virology*, Vol. 16 (H. Fraenkel-Conrat and R. R. Wagner, eds.), p. 257, Plenum Press, New York.

44b. Manuelidis, L., and Manuelidis, E. E., 1981, Search for specific DNAs in Creutzfeldt–Jakob infectious brain fractions using "nick translation," *Virology* **109:**435.

44c. Gajdusek, D. C., 1977, Unconventional viruses and the origin and disappearance of kuru, *Science* **197:**943.

45a. Eaton, B. T., and Steacie, A. D., 1980, Cricket paralysis virus RNA has a 3' terminal poly(A), *J. Gen. Virol.* **50:**167.

45b. Scotti, P. D., and Longworth, J. F., 1980, The biology and ecology of strains of an insect small RNA virus complex, *Adv. Virus Res.* **26:**117.

46a. Matthews, R. E. F., 1982, Classification and nomenclature of viruses, *Intervirology* **17:**49.

46b. Rapp, F., 1980, Persistence and transmission of cytomegalovirus, in: *Comprehensive Virology*, Vol. 16 (H. Fraenkel-Conrat and R. R. Wagner, eds.), p. 193, Plenum Press, New York.

46c. Stinski, M. F., 1983, Molecular biology of cytomegaloviruses, in: *The Viruses*, The Herpesviruses, Vol. II (H. Fraenkel-Conrat, R. R. Wagner, and B. Roizman, eds.), p. 67, Plenum Press, New York.

46d. Rapp, F. 1983, The biology of cytomegaloviruses, in: *The Viruses*, The Herpesviruses, Vol. II (H. Fraenkel-Conrat, R. R. Wagner, and B. Roizman, eds.), p. 1, Plenum Press, New York.

47a. Matthews, R. E. F., 1982, Classification and nomenclature of viruses, *Intervirology* **17:**86.

47b. Payne, C. C., and Mertens, P. P. C., 1983, Cytoplasmic polyhedrosis viruses, in: *The Viruses*, The Reoviridae (H. Fraenkel-Conrat, R. R. Wagner, and B. Roizman, eds.), p. 425, Plenum Press, New York.

48a. Kelly, D. C., and Bud, H. M., 1978, Densonucleosis virus DNA: Analysis of fine structure by electron microscopy and agarose gel electrophoresis, *J. Gen. Virol.* **40:**33.

48b. Tijssen, P., and Kurstak, E., 1981, Biochemical, biophysical, and biological properties of densonucleosis virus (parvovirus). III. Common sequences of structural proteins, *J. Virol.* **37:**17–23.

48c. Matthews, R. E. F., 1982, Classification and nomenclature of viruses, *Intervirology* **17:**74.

49a. Teninges, D., 1979, Protein and RNA composition of the structural components of *Drosophila* X virus, *J. Gen. Virol.* **45:**641.

49b. Teninges, D., Ohanessian, A., Richard-Molard, C., and Contamine, D., 1979, Isolation and biological properties of *Drosophila* X virus, *J. Gen. Virol.* **42:**241.

49c. Scott, M. P., Fostel, J. M., and Pardue, M. L., 1980, A new type of virus from cultured *Drosophila* cells: Characterization and use in studies of the heat-shock response, *Cell* **22:**929.

50. Tignor, G. H., Murphy, F. A., Clark, H. F., Shope, R. E., Madore, P., Bauer, S. P., Buckley, S. M., and Meredith, D. C., 1977, Duvenhage virus: Morphological, biochemical, histopathological and antigenic relationships to the rabies serogroup, *J. Gen. Virol.* **37:**595.

51. Rosenwirth, B., and Eggers, H. J., 1978, Structure and replication of echovirus type

12. 1. Analysis of the polypeptides and RNA of echovirus 12 particles, *Eur. J. Biochem.* **92**:53.

52. Todd, D., and McNulty, M. S., 1978, Biochemical studies on a virus associated with egg drop syndrome 1976, *J. Gen. Virol.* **40**:63.

53b. Olberding, K. P., and Frost, J. W., 1975, Electron microscopical observations of the structure of the virus of viral haemorrhagic septicaemia (VHS) of rainbow trout (*Salmo gairdneri*), *J. Gen. Virol.* **27**:305.

54. Merregaert, J., van Emmelo, J., Devos, R., Porter, Al., Fellner, P., and Fiers, W., 1978, The 3'-terminal nucleotide sequence of encephalomyocarditis virus RNA, *Eur. J. Biochem.* **82**:55.

56. Matthews, R. E. F., 1982, The entomopoxvirinae, *Intervirology* **17**:44.

56a. Kieff, E., Dambaugh, T., King, W., Heller, M., Cheung, A., van Santen, V., Hummel, M., Beisel, C., and Fennewald, S., 1983, Biochemistry of Epstein–Barr virus, in: *The Viruses*, The Herpesviruses, Vol. I (H. Fraenkel-Conrat, R. R. Wagner, and B. Roizman, eds.), p. 105, Plenum Press, New York.

56b. Henle, W., and Henle, G., 1983, Immunology of Epstein–Barr virus, in: *The Viruses*, The Herpesviruses, Vol. I (H. Fraenkel-Conrat, R. R. Wagner, and B. Roizman, eds.), p. 200, Plenum Press, New York.

56c. de-Thé, G., 1983, Epidemiology of Epstein–Barr virus and associated diseases in man, in: *The Viruses*, The Herpesviruses, Vol. I (H. Fraenkel-Conrat, R. R. Wagner, and B. Roizman, eds.), p. 25, Plenum Press, New York.

57. Zeegers, J. J. W., Van der Zeljst, B. A. M., and Horznnek, M. C., 1976, The structural proteins of equine arteritis virus, *Virology* **73**:200.

58. O'Callaghan, D. J., Gentry, G. A., and Randall, C. C., 1983, The equine herpesviruses, in: *The Viruses*, The Herpesviruses, Vol. II (H. Fraenkel-Conrat, R. R. Wagner, and B. Roizman, eds.), p. 215, Plenum Press, New York.

59a. Parekh, B., Issel, C. J., and Montelaro, R. C., 1980, Equine infectious anemia virus, a putative lentivirus, contains polypeptides analogous to prototype-C oncornaviruses, *Virology* **107**:520–525.

59b. Summers J., Jones, S. E., and Anderson, M. J., 1983, Characterisation of the agent of erythrocyte aplasia permits its classification as a human parvovirus, *J. Gen. Virol.* **64**:2527.

60. Clark, H. F., Lief, F. S., Lunger, P. D., Waters, D., Leloup, P., Foelsch, D. W., and Wyler, R. W., 1979, Fer de Lance virus (FDLV): A probable paramyxovirus isolated from a reptile, *J. Gen. Virol.* **44**:405–418.

61. Kiley, M. P. *et al.*, 1983, A taxonomic home for marburg and ebolavirus, *Intervirology* **18**:24.

62. Matthews, R. E. F., 1982, Classification and nomenclature of viruses, *Intervirology* **17**:98.

63a. Goorha, R., and Murti, K. G., 1982, The genome of frog virus 3, an animal DNA virus, is circularly permuted and terminally redundant, *Proc. Natl. Acad. Sci. USA* **79**:248–252.

63b. Murti, K. G., Goorha, R., and Granoff, A., 1982, Structure of frog virus 3 genome: Size and arrangement of nucleotid sequences as determined by electron microscopy, *Virology* **116**:275–283.

64. Matthews, R. E. F., 1982, Classification and nomenclature of viruses, *Intervirology* **17**:50.

65. Ganem, D., Greenbaum, L., and Varmus, H. E., 1982, Virion DNA of ground squirrel hepatitis virus: Structural analysis and molecular cloning, *J. Virol.* **44**:374–383.

66. Koller, R., and Goulian, M., 1981, Synthesis of parvovirus H-1 replicative form from viral DNA by DNA polymerase γ, *Proc. Natl. Acad. Sci. USA* **78**:6206–6210.

67. Foulke, R. S., Rosato, R. R., and French, G. R., 1981, Structural polypeptides of hazara virus, *J. Gen. Virol.* **53**:169–172.

68. Bosslet, K., and Sauer, G., 1978, Biological properties and physical map of the genome of a new papovavirus, HD virus, *J. Virol.* **25**:596–607.

69. Pocock, D. H., 1978, Effect of sulphydryl reagents on the biological activities, polypeptide composition and morphology of haemagglutinating encephalomyelitis virus, *J. Gen. Virol.* **40**:93–101.

70. Gust, I. D., Coulepis, A. G., Feinstone, S. M., Locarnini, S. A., Moritsugu, Y., Najera, R., and Siegl, G., 1983, Taxonomic classification of hepatitis A virus, *Intervirology* **20**:1–7.

71a. Robinson, W. S., Viruses of human hepatitis A and B, in: *Comprehensive Virology*, Vol. 14 (H. Fraenkel-Conrat and R. R. Wagner, eds.), pp. 471–526, Plenum Press, New York.

71b. Melnick, j. L., 1983, Class of hepatitis A virus as entero type 72 and hepatitis B as hepadnavirus Type 1, *Intervirology* **18**:103.

72a. Galibert, F., Mandart, E., Fitoussi, F., Tiollais, P., and Charnay, P., 1979, Nucleotide sequence of the hepatitis B virus genome (subtype ayw) cloned in *E. coli*, *Nature* **281**:646.

72b. Pasek, M., Goto, T., Gilbert, W., Zink, B., Schaller, H., MacKay, P., Leadbetter, Gl., and Murray, K., 1979, Hepatitis B virus genes and their expression in *E. coli*, *Nature* **282**:575.

73. Fleckenstein, B., and Desrosiers, R. C., 1983, *Herpesvirus saimiri* and *Herpesvirus ateles*, in: *The Viruses*, The Herpesviruses, Vol. I (H. Fraenkel-Conrat, R. R. Wagner, and B. Roizman, eds.), p. 253, Plenum Press, New York.

74a. Roizman, B., 1983, The familey Herpesviridae: General description, taxonomy, and classification, in: *The Viruses*, The Herpesviruses, Vol. I (H. Fraenkel-Conrat, R. R. Wagner, and B. Roizman, eds.), p. 1, Plenum Press, New York.

74b. Matthews, R. E. F., 1982, Classification and nomenclature of viruses, *Intervirology* **17**:47.

75. Krell, P. J., and Stoltz, D. B., 1980, Virus-like particles in ovary of an ichneumonid wasp: Purification and preliminary characterization, *Virology* **101**:408.

76. Robert-Guroff, M., Fahey, K. A., Maeda, M., Nakao, Y., Ito, Y., and Gallo, R. C., 1982, Identification of HTLV p19 specific natural human antibodies by competition with monoclonal antibody, *Virology* **122**:297.

77. Langridge, W. H. R., 1981, Biochemical properties of a persistent nonoccluded baculovirus isolated from *Heliothis zea* cells, *Virology* **112**:770.

78a. Compans, R. W., and Choppin, P. W., 1975, Reproduction of povaviruses, in: *Comprehensive Virology*, Vol. 4 (H. Fraenkel-Conrat and R. R. Wagner, eds.), p. 179, Plenum Press, New York.

78b. Laver, W. G., Air, G. M., Webster, R. G., and Markoff, L. J., 1982, Amino acid sequence changes in antigenic variants of type A influenza virus N2 neuraminidase, *Virology* **122**:450.

78c. Webster, R. G., Laver, W. G., Air, G. M., and Schild, G. C., 1982, Molecular mechanisms of variation in influenza viruses, *Nature* **296**:115.

78d. Kaptein, J. S., and Nayak, D. P., 1982, Complete nucleotide sequence of the polymerase 3 gene of human influenza virus A/WNS/33, *J. Virol.* **42**:55.

78e. Winter, G., and Fields, S., 1982, Nucleotide sequence of human influenza A/PR/8/34 segment 2, *Nucl. Acids Res.* **10**:2135.

79. Matthews, R. E. F., 1982, Classification and nomenclature of viruses, *Intervirology* **17**:56.

80. Huang, C. H., 1982, Studies of Japanese encephalitis in China, *Adv. Virus Res.* **27**:71.

81a. Martin, J. D., Padgett, B. L., and Walker, D. L., 1983, Characterization of tissue culture-induced heterogeneity in DNAs of independent isolates of JC virus, *J. Gen. Virol.* **64**:2271.

81b. Walker, D. L., and Padgett, B. L., 1983, Progressive multifocal leukoencephalopathy, in: *Comprehensive Virology*, Vol. 18 (H. Fraenkel-Conrat and R. R. Wagner, eds.), p. 161, Plenum Press, New York.

82. Siddell, S. G., 1982, Coronavirus JHM: Tryptic peptide fingerprinting of virion proteins and intracellular polypeptides, *J. Gen. Virol.* **62**:259.

83. Pudney, M., Newman, J. F. E., and Brown, F., 1978, Characterization of kawino virus, an entero-like virus isolated from the mosquito *Mansonia uniformis* (diptera: culicidae), *J. Gen. Virol.* **40**:433.

84. Scotti, P. D., Gibbs, A. J., and Wrigley, N. G., 1976, Kelp fly virus, *J. Gen. Virol.* **30**:1.

85. Westaway, E. G., and Shew, M., 1977, Proteins and glycoproteins specified by the flavivirus kunjin, *Virology* **80**:309.

86. Gentsch, J., Wynne, L. R., Clewley, J. P., Shope, R. E., and Bishop, D. H. L., 1977, Formation of recombinants between snowshoe hare and la crosse bunyaviruses, *J. Virol.* **24**:893.

87. Matthews, R. E. F., 1982, Classification and nomenclature of viruses, *Intervirology* **17**:127.

88a. Lehmann-Grube, F., Peralta, L. M., Bruns, M., and Lohler, J., 1983, Persistent infection of mice with the lymphocytic choriomeningitis virus, in: *Comprehensive Virology*, Vol. 18 (H. Fraenkel-Conrat and R. R. Wagner, eds.), p. 43, Plenum Press, New York.

88b. Oldstone, M. B. A., 1979, Immune responses, immune tolerance, and viruses, in: *Comprehensive Virology*, Vol. 15 (H. Fraenkel-Conrat and R. R. Wagner, eds.), p. 1, Plenum Press, New York.

88c. Buchmeier, M. J., Lewicki, H. A., Tomori, O., and Johnson, K. M., 1980, Monoclonal antibodies to lymphocytic choriomeningitis virus react with pathogenic arenaviruses, *Nature* **288**:486.

89. Matthews, R. E. F., 1982, Classification and nomenclature of viruses, *Intervirology* **17**:111.

90. Nonoyama, M., 1983, The molecular biology of Marek's disease herpesvirus, in: *The Viruses*, The Herpesviruses, Vol. I (H. Fraenkel-Conrat, R. R. Wagner, and B. Roizman, eds.), p. 333, Plenum Press, New York.

91. Matthews, R. E. F., 1982, The mastadenoviruses, *Intervirology* **17**:60.

92. Perez-Bercoff, R., and Gander, M., 1977. The genomic RNA of mengovirus. I. Location of the poly(C) tract, *Virology* **80**:426.

93a. Chattopadhyay, S. K., Cloyd, M. W., Linemeyer, D. L., Lander, M. R., Rands, E., and Lowy, D. R., 1982, Cellular origin and role of mink cell focus-forming viruses in murine thymic lymphomas, *Nature* **295**:25.

93b. Chattopadhyay, S. K., Lander, M. R., Gupta, S., Rands, E., and Lowy, D. R., 1981, Origin of mink cytopathic focus-forming (MCF) viruses: Comparison with ecotropic and xenotropic murine leukemia virus genomes, *Virology* **113**:465.

94. Cotmore, S. F., Sturzenbecker, L. J., and Tattersall, P., 1983, The autonomous parvovirus MVM encodes two nonstructural proteins in addition to its capsid polypeptides, *Virology* **129**:333.

95. Matthews, R. E. F., 1978, Classification and nomenclature of viruses, *Intervirology* **17**:105.

96. Stohlman, S. A., and Lai, M. M. C., 1979, Phosphoproteins of murine hepatitis viruses, *J. Virol.* **22**:672.

97a. Yagi, M. J., and Compans, R. W., 1977, Structural components of mouse mammary tumor virus. I. Polypeptides of the virion, *Virology* **76**:751.

97b. Ball, J. K., Dekaban, G. A., McCarter, J. A., and Loosmore, S. M., 1983, Molecular biological characterization of a highly leukaemogenic virus isolated from the mouse. III. Identity with mouse mammary tumour virus, *J. Gen. Virol.* **64**:2177.

98. Örvell, C., 1978, Structural polypeptides of mumps virus, *J. Gen. Virol.* **41**:527.

99a. Matthews, R. E. F., 1982, Classification and nomenclature of viruses, *Intervirology* **17**:117.

99b. Casals, J., and Tignor, G. H., 1979, The nairovirus genus; serological interrelationships, *Intervirology* **14**:144.

100. Newman, J. F. E., and Brown, F., 1977, Further physicochemical characterization of nodamura virus. Evidence that the divided genome occurs in a single component, *J. Gen. Virol.* **38**:83.

101. Matthews, R. E. F., 1982, The Nodaviridae, *Intervirology* **17**:167.

102a. Matthews, R. E. F., 1982, Nudaurelia β viruses, *Intervirology* **17**:135.

102b. Juckes, I. R. M., 1979, Comparison of some biophysical properties of the nudaurelia β and ε viruses, *J. Gen. Virol.* **42**:89–94.

102c. Reinganum, C., Robertson, J. S., and Tinsley, T. W., 1978, A new group of RNA viruses from insects, *J. Gen. Virol.* **40**:195–202.

103a. Matthews, R. E. F., 1982, The Oncovirinae, *Intervirology* **17**:125.

103b. Vogt, P. K., 1977, Genetics of RNA tumor viruses, in: *Comprehensive Virology*, Vol. 9 (H. Fraenkel-Conrat and R. R. Wagner, eds.), p. 341, Plenum Press, New York.

103c. Hanafusa, H., 1977, Cell transformation by RNA tumor viruses, in: *Comprehensive Virology*, Vol. 10 (H. Fraenkel-Conrat and R. R. Wagner, eds.), p. 401, Plenum Press, New York.

103d. Duesberg, P., 1979, Transforming genes of retroviruses, Cold Spring Harbor Symposium on Quantitative Biology **45**:13.

104a. Verwoerd, D. W., Huismans, H., and Erasmus, B. J., 1979, Orbiviruses, in: *Comprehensive Virology*, Vol. 14 (H. Fraenkel-Conrat and R. R. Wagner, eds.), p. 285, Plenum Press, New York.

104b. Gorman, B. M., Taylor, J., and Walker, P. J., 1983, Orbiviruses, in: *The Viruses* The Reoviridae (H. Fraenkel-Conrat, R. R. Wagner, and B. Roizman, eds.), p. 287, Plenum Press, New York.

104c. Matthews, R. E. F., 1982, The Orbiviruses, *Intervirology* **17**:82.

105a. Matthews, R. E. F., 1982, The Orthomyxoviridae, *Intervirology* **17**:106.

105b. Hightower, L. E., and Bratt, M. A., 1975, Genetics of orthomyxoviruses, in: *Comprehensive Virology*, Vol. 9 (H. Fraenkel-Conrat and R. R. Wagner, eds.), p. 535, Plenum Press, New York.

106. Matthews, R. E. F., 1982, The orthopoxviruses, *Intervirology* **17**:42.

107. Matthews, R. E. F., 1982, The orthoreoviruses, *Intervirology* **17**:81.

108. Payne, C. C., 1974, The isolation and characterization of a virus from *Oryctes rhinoceros*, *J. Gen. Virol.* **25**:105.

109. Matthews, R. E. F., 1982, Classification and nomenclature of viruses, *Intervirology* **17**:62.

110a. Salzman, N. P., and Khoury, G., 1979 Reproduction of papovaviruses, in: *Comprehensive Virology*, Vol. 3 (H. Fraenkel-Conrat and R. R. Wagner, eds.), p. 63, Plenum Press, New York.

110b. Finch, J. T., and Crawford, L. V., 1975, Structure of small DNA-containing animal viruses, in: *Comprehensive Virology*, Vol. 4 (H. Fraenkel-Conrat and R. R. Wagner, eds.), p. 119, Plenum Press, New York.

110c. Matthews, R. E. F., 1982, Classification and nomenclature of viruses, *Intervirology* **17**:62.

111a. Choppin, P. W., and Compans, R. W., 1975, Reproduction of paramyxoviruses, in: *Comprehensive Virology*, Vol. 4 (H. Fraenkel-Conrat and R. R. Wagner, eds.), p. 95, Plenum Press, New York.

111b. Matthews, R. E. F., 1982, Classification and nomenclature of viruses, *Intervirology* **17**:104.

111c. Kingsbury, D. W., Bratt, M. A., Choppin, W., Hanson, R. P., Hosaka, Y., ter Muelen, V., Norrby, E., Plowright, W., Rott, R., and Wunner, W. H., 1978, Paramyxoviridae, *Intervirology* **10**:137.

111d. Matthews, H. E. F., 1982, Classification and nomenclature of viruses, *Intervirology* **17**:43.

112a. Rose, J. A., 1974, Parvovirus reproduction, in: *Comprehensive Virology*, Vol. 3 (H. Fraenkel-Conrat and R. R. Wagner, eds.), p. 1, Plenum Press, New York.

112b. Bachmann, P. A., Hoggan, M. D., Kurstak, E., Melnikc, J. L., Pereira, H. G., Tattersall, P., and Vago, C., 1979, Parvoviridae: Second report, *Intervirology* **11**:248.

112c. Matthews, H. E. F., 1982, Classification and nomenclature of viruses, *Intervirology* **17**:72.

113. Matthews, H. E. F. 1982, Classification and nomenclature of viruses, *Intervirology* **17**:100.

114. Matthews, H. E. F., 1982, Classification and nomenclature of viruses, *Intervirology* **17:**116.

115a. Crowell, R. L., and Landau, B. J., 1983, Receptors in the initiation of picornavirus infections, in: *Comprehensive Virology*, Vol. 18 (H. Fraenkel-Conrat and R. R. Wagner, eds.), p. 1, Plenum Press, New York.

115b. Rueckert, R. R., 1975, On the structure and morphogenesis of picornaviruses, in: *Comprehensive Virology*, Vol. 5 (H. Fraenkel-Conrat and R. R. Wagner, eds.), p. 31, Plenum Press, New York.

115c. Cooper, P. D., 1977, Genetics of picornaviruses, in: *Comprehensive Virology*, Vol. 9 (H. Fraenkel-Conrat and R. R. Wagner, eds.), p. 133, Plenum Press, New York.

115d. Cooper, P. D., Agol, V. I., Bachrach, H. L., Brown, F., Ghendon, Y., Gibbs, A. J., Gillespie, J. H., Lonberg-Holm, K., Mandel, B., Melnick, J. L., Mohanty, S. B., Povey, R. C., Rueckert, R. R., Schaffer, F. L., and Tyrrell, D. A. J., 1978, Picornaviridae: Second report, *Intervirology* **10:**165.

115e. Matthews, R. E. F., 1982, Classification and nomenclature of viruses, *Intervirology* **17:**129.

116. Matthews, R. E. F., 1982, Classification and nomenclature of viruses, *Intervirology* **17:**105.

117. Kitamura, N., Semler, B. L., Rothberg, P. G., Larsen, G. R., Adler, C. J., Dorner, A. J., Emini, E. A., Hanecak, R., Lee, J. J., van der Werf, S., Anderson, C. W., and Wimmer, E., 1981, Primary structure, gene organization and polypeptide expression of poliovirus RNA, *Nature* **291:**547.

117a. Stoltz *et al.*, 1984, Polydnaviridae: proposed family of insect viruses with segmented double-stranded circular DNA genome, *Intervirology* **21:**1.

118a. Eckhart, W., 1977, Genetics of polyoma virus and simian virus 40, in: *Comprehensive Virology*, Vol. 9 (H. Fraenkel-Conrat and R. R. Wagner, eds.), p. 1, Plenum Press, New York.

118b. Matthews, R. E. F., 1982, Classification and nomenclature of viruses, *Intervirology* **17:**62.

119a. Moss, B., 1974, Reproduction of poxviruses, in: *Comprehensive Virology*, Vol. 3 (H. Fraenkel-Conrat and R. R. Wagner, eds.), p. 405, Plenum Press, New York.

119b. Gafford, L. G., Mitchell, E. B., Jr., and Randall, C. C., 1978, Sedimentation characteristics and molecular weights of three poxvirus DNAs, *Virology* **89:**229.

119c. Matthews, R. E. F., 1982, Classification and nomenclature of viruses, *Intervirology* **17:**42.

120. Matthews, R. E. F., 1982, Classification and nomenclature of viruses, *Intervirology* **17:**57.

121a. Sharpe, A. H., and Fields, B. N., 1983, Pathogenesis of reovirus infection, in: *The Viruses*, The Reoviridae (H. Fraenkel-Conrat, R. R. Wagner, and B. Roizman, eds.), p. 229, Plenum Press, New York.

121b. Joklik, W. K., 1983, *The Viruses*, The Reoviridae (H. Fraenkel-Conrat and R. R. Wagner, eds.), p. 1, Plenum Press, New York.

121c. Joklik, W. K., 1983, The reovirus particle, in: *The Viruses*, The Reoviridae (H. Fraenkel-Conrat and R. R. Wagner, eds.), p. 9, Plenum Press, New York.

121d. Shatkin, A. J., and Kozak, M., 1983, Biochemical aspects of reovirus transcription and translation, in: *The Viruses*, The Reoviridae (H. Fraenkel-Conrat and R. R. Wagner, eds.), p. 79, Plenum Press, New York.

121e. Matthews, R. E. F., 1982, Classification and nomenclature of viruses, *Intervirology* **17:**81.

122. Lambert, D. M., Pons, M. W., Mbuy, G. N., and Hasler, K. D., 1980, Nucleic acids of respiratory syncytial virus, *J. Virol.* **36:**837.

123a. Matthews, R. E. H., 1982, Classification and nomenclature of viruses, *Intervirology* **17:**124.

123b. Duesberg, P. H., 1983, Retroviral transforming genes in normal cells? *Nature* **304:**219.

124a. Wagner, R. R., 1975, Reproduction of rhabdoviruses, in: *Comprehensive Virology*, Vol. 4 (H. Fraenkel-Conrat and R. R. Wagner, eds.), p. 1, Plenum Press, New York.

124b. Pringle, C. R., 1977, Genetics of rhabdoviruses, in: *Comprehensive Virology*, Vol. 9 (H. Fraenkel-Conrat and R. R. Wagner, eds.), p. 239, Plenum Press, New York.

124c. Matthews, R. E. F., 1982, Classification and nomenclature of viruses, *Intervirology* **17:**109

125. Matthews, R. E. F., 1982, Classification and nomenclature of viruses, *Intervirology* **17:**130.

126a. Holmes, I. H., 1983, Rotaviruses, in: *The Viruses*, The Reoviridae (H. Fraenkel-Conrat, R. R. Wagner, and B. Roizman, eds.), p. 359, Plenum Press, New York.

126b. Matthews, R. E. F., 1982, Classification and nomenclature of viruses, *Intervirology* **17:**84.

127a. Keith, J., and Fraenkel-Conrat, H., 1975, Identification of the 5' end of rous sarcoma virus RNA, *Proc. Natl. Acad. Sci. USA* **72:**3347.

127b. Hackett, P. B., Swanstrom, R., Varmus, H. E., and Bishop, J. M., 1982, The leader sequence of the subgenomic mRNA's of rous sarcoma virus is approximately 390 nucleotides, *J. Virol.* **41:**527.

128. Van Alstyne, D., Drystal, G., Kettyls, G. D., and Bohn, E. M., 1981, The purification of rubella virus (RV) and determination of its polypeptide composition, *Virology* **108:**491.

129a. Marsh, R. F., Malone, T. G., and Lancaster, R. P., 1978, Evidence for an essential DNA component in the scrapie agent, *Nature* **275:**147.

129b. Cho, H. J., 1979, Requirement of a protein component for scrapie infection, *Intervirology* **14:**213.

130. Monroe, S. S., Ou J-H., Rice, C. M., Schlesinger, S., Strauss, E. G., and Strauss, J. H., 1982, Sequence analysis of cDNAs derived from the RNA of sindbis virions and of defective interfering particles, *J. Virol.* **41:**153.

131a. Schultz, A. M., Ruscetti, S. K., Scolnick, E. M., and Oroszlan, S., 1980, The *env*-gene of the spleen focus-forming virus lacks expression of p15(E) determinants, *Virology* **107:**537.

131b. Linemeyer, D. L., Ruscetti, S. K., Scolnick, E. M., Evans, L. H., and Duesberg, P. H., 1981, Biological activity of the spleen focus-forming virus is encoded by a molecularly cloned subgenomic fragment of spleen focus-forming virus DNA, *Proc. Natl. Acad. Sci. USA* **78:**1401.

132. O'Rear, J. J., and Temin, H. M., 1981, Mapping of alterations of noninfectious proviruses of spleen necrosis virus, *J. Virol.* **39:**138.

133. Matthews, R. E. F., 1982, Classification and nomencalture of viruses, *Intervirology* **17:**126.

134. ter Meulen, V., Stephenson, J. R., and Kreth, H. W., 1974, Subacute sclerosing panencephalitis, in: *Comprehensive Virology*, Vol. 18 (H. Fraenkel-Conrat and R. R. Wagner, eds.), p. 105, Plenum Press, New York.

135. Matthews, R. E. F., 1982, Classification and nomenclature of viruses, *Intervirology* **17:**44.

136. Fiers, W., Contreras, R., Haegeman, G., Rogiers, R., Van de Voorde, A., Van Heuverswyn, H., Van Herreweghe, V., Volckaert, G., and Ysebaert, M., 1978, Complete nucleotide sequence of SV40 DNA, *Nature* **273:**113.

137a. Matthews, R. E. F., 1982, Classification and nomenclature of viruses, *Intervirology* **17:**97.

137b. Pfefferkorn, E. R., and Shapiro, D., 1974, Reproduction of togaviruses, in: *Comprehensive Virology*, Vol. 2 (H. Fraenkel-Conrat and R. R. Wagner, eds.), p. 171, Plenum Press, New York.

137c. Pfefferkorn, E. R., 1977, Genetics of togaviruses, in: *Comprehensive Virology*, Vol. 9 (H. Fraenkel-Conrat and R. R. Wagner, eds.), p. 209, Plenum Press, New York.

138a. Matthews, R. E. F., 1982, Classification and nomenclature of viruses, *Intervirology* **17:**117.

138b. Hewlett, M. J., Pettersson, R. F., and Baltimore, D., 1977, Circular forms of uukuniemi virion RNA: An electron microscopic study, *J. Virol.* **21:**1085.

139a. Belle Isle, H., Venkatesan, S., and Moss, B., 1981, Cell-free translation of early and late mRNAs selected by hybridization to cloned DNA fragments derived from the left 14-million to 72-million daltons of the vaccinia virus genome, *Virology* **112:**306.

139b. Wittek, R., Barbosa, E., Cooper, J. A., Garon, C. F., Chan, H., and Moss, B., 1980, Inverted terminal repetition in vaccinia virus DNA encodes early mRNAs, *Nature* **285:**21.

140. Hyman, R. W., 1983, Molecular biology of varicella–zoster virus, in: *The Viruses*, The Herpesviruses, Vol. II (H. Fraenkel-Conrat, R. R. Wagner, and B. Roizman, eds.), p. 115, Plenum Press, New York.

141. Matthews, R. E. F., 1982, Classification and nomenclature of viruses, *Intervirology* **17:**110.

142. Harris, J. D., Scott, J. V., Traynor, B., Brahic, M., Stowring, L., Ventura, P., Haase, A. T., and Peluso, R., 1981, Visna virus DNA: Discovery of a novel gapped structure, *Virology* **113:**573.

143a. DeJean, A., Vitvitski, L., Brechot, C., Trepo, C., Tiollais, P., and Charnay, P., 1982, Presence and state of woodchuck hepatitis virus DNA in liver and serum of woodchucks: Further analogies with human hepatitis B virus, *Virology* **121:**195.

143b. Galibert, F., Chen, T. N., and Mandart, E., 1982, Nucleotide sequence of a cloned woodchuck hepatitis virus genome: Comparison with the hepatitis B virus sequence, *J. Virol.* **41:**51.

Plant Viruses, Including Protophytal Viruses

Those viruses that have been identified as belonging to one of the groups approved by the ICTV will only be identified in that manner. Specific properties described in the literature, particularly numerical ones such as molecular weights, may not agree exactly with those given for the type member or the general range describing the group. However, in view of the uncertainty of all such values (except for sequenced molecules), only gross deviations from group characteristics will be noted for individual viruses.

Viruses that do not fit into any group in biological and/or chemical respects will be termed unclassified, and their main properties listed. Certain well-characterized viruses, such as alfalfa mosaic and tobacco necrosis viruses, appear to be singular, and will be termed monotypic and will be described as such. They may become type members of new groups.

It should be realized that many classifications are made on dubious grounds, particularly among the filamentous viruses, and that many viruses, notwithstanding isolation from different hosts and carrying different names, may actually be the same.

Most plant viruses contain single-stranded plus-strand RNA, and only those with different types of genomes will be so identified (Reoviridae, mycoviruses, Rhabdoviridae, and the single- or double-stranded DNA-containing viruses, the gemini and caulimo virus groups. Most viruses are transmissible or are naturally transmitted by sap, also termed mechanically, and only those not so transmissible will be identified. Host ranges will be termed narrow or wide depending on how many different plant families, genera, or species have been shown or found to become infected by them.

Virus	Classification or properties (Group names in Capitals)
Abaca mosaic	POTYVIRUS (serologically related to sugarcane mosaic virus)
Af-S (*Aspergillus foetidus*)	MYCOVIRUS (double-stranded RNA)
Agropyron mosaic	POTYVIRUS (15 × 717 nm particles, mite-transmitted with *Graminae* as only hosts, serologically distantly related to hordeum mosaic and wheat streak mosaic virus).
Alfalfa 1 and 2	synonym for alfalfa mosaic virus
Alfalfa latent	CARLAVIRUS (635 nm), (no serological relationship detected)
ALFALFA MOSAIC	Monotypic: Composed of four types of bacilliform particles, 18 × 58, 48, 36, and 28 nm (B, M, Tb, Ta; 94 S, 80 S, 73 S, 66 S) (Figure 18). Their density in Cs_2SO_4 is about 1.28 g/cm^3. These contain four RNAs (RNA 1–4) of 1.04, 0.73, 0.62, and 0.28 × 10^6 daltons (about 16% of particle weight). The particle weights are 6.9, 5.2, 4.3, and 3.8 × 10^6. For infectivity, all four RNAs or the three larger ones plus some viral coat protein are required. The coat protein of 220 amino acids has a molecular weight of 24,280.

The RNAs carry caps (7-MeG$^{5'}$-ppp$^{5'}$Gp-) at the 5' end and no 3' terminal poly(A). RNA 4 has been sequenced for strain 425. Its sequence occurs also at the 3' end of RNA 3. RNA 1 and 2 are monocistronic for proteins of 100 and 80 × 10^3 daltons. RNA 3 carries the gene for a third protein of 35 × 10^3 daltons, the gene for the coat protein being silent in RNA 3, and expressed in RNA 4.

The virus infects many hosts, often without symptoms; it is transmitted in nonpersistent manner by aphids, but also by sap and seed transmission. Several strains differing in host preference, symptomology, and/or chemical detail have been studied.

Virus	Classification or properties (Group names in Capitals)
Algal viruses[2]	Virus or virus-like particles (VLP) have been detected in many algal species. Most have not been named or well characterized, and will not be included in this catalogue. Included are the *Chara* viruses which resemble tobamo viruses, and the viruses isolated from chlorella-like algae that occur symbiotically in *Paramecium bursaria* and *Hydra viridis* (PBCV-1, HVCV-1, and HVCV-2). The large viruses of about 190 nm diameter (2500 S) carry DNA (about 150×10^6 daltons).
Alliaria mosaic	POTYVIRUS or POTEXVIRUS
Almovirus group	name proposed for alfalfa mosaic virus group
Amaranthus leaf mottle	POTYVIRUS
American wheat striate mosaic: *see* wheat striate mosaic	
Amyelois chronic stunt	possibly member of CALICIVIRIDAE
Andean potato latent	TYMOVIRUS (possibly a strain of eggplant mosaic virus)
Andean potato mottle	COMOVIRUS (distantly related to other comoviruses)
Annulus tabaci	synonym for tobacco ring spot
Annulus zonatus	synonym for tomato ring spot
An-S (*Aspergillus niger*)	MYCOVIRUS (double-stranded RNA)
Anthoxanthum mosaic	possibly potyvirus
Anthriscus yellows	helper of parsnip yellow fleck virus for aphid transmission
Apple chlorotic leaf spot	CLOSTEROVIRUS (subgroup II), (12×730 nm labile threads)
Apple infectious variegation	synonym for apple mosaic virus
Apple latent 2	synonym for sowbane mosaic
Apple mosaic	ILARVIRUS (possibly identical with Danish plum line pattern, hop A, and rose mosaic)

Virus	Classification or properties (Group names in Capitals)
Apple stem grooving	unclassified (12 × 619 nm, no known vectors, no serological relationships discovered); possibly member of proposed capillovirus group
Apple (Tulare) mosaic: *see* Tulare apple mosaic	
Aquilegia	possibly potyvirus
Arabis mosaic	NEPOVIRUS
Aranjia mosaic	possibly potyvirus
Arracacha A[3]	probably nepovirus (no known vectors, no serological relationship)
Arracacha B[4]	unclassified 26-nm-diameter particles with two coat proteins, 126 S, 40% RNA
Arrhenatherum blue dwarf	FIJI VIRUS (REOVIRIDAE), (identical to oat sterile dwarf and lolium enation virus)
Artichoke curly dwarf	possibly potexvirus
Artichoke Italian latent	NEPOVIRUS (no serological relationship to other nepoviruses)
Artichoke mottled crinkle	TOMBUSVIRUS
Artichoke vein banding	possibly nepovirus
Artichoke yellow ringspot	NEPOVIRUS
Aspergillus spp.	MYCOVIRUSES (double-stranded RNA)
Aster ringspot	strain of tobacco rattle virus
Asystasia mosaic	possibly potyvirus
Atopa mild mosaic	synonym or strain of henbane mosaic virus
Aucuba mosaic (yellow and green)	closely related strains of tobacco mosaic virus
Avocado sunblotch	VIROID (only 274 nucleotide circles)
Azuki bean mosaic	synonym for cowpea aphid-borne mosaic virus
Bambarra groundnut	CARLAVIRUS (aphid-transmitted)
Bamboo mosaic	possibly potexvirus

Virus	Classification or properties (Group names in Capitals)
Banana bunchy top	possibly luteovirus
Barley B-1	possibly potexvirus
Barley (false) stripe mosaic (BSMV)[5]	type member of HORDEIVIRUS group
Barley yellow dwarf (BYDV)[6]	type member of LUTEOVIRUS group
Barley yellow mosaic	possibly potyvirus (fungus-transmitted by *Polymyxa graminis*)
Barley yellow striate mosaic	PHYTORHABDOVIRUS (RHABDOVIRIDAE), (hopper-transmitted)
Barrel cactus	POTEXVIRUS
Bean 1	synonym for bean common mosaic
Bean (common) mosaic	POTYVIRUS
Bean curly dwarf mosaic	COMOVIRUS
Bean golden mosaic	type member of GEMINIVIRUS (single-stranded DNA)
Bean leaf roll	identical with pea leaf roll
Bean mild mosaic[7]	unclassified (isomeric 28 nm diameter, 175 S particles, 20% RNA of 1.3×10^6 daltons, serologically not related to any virus tested, transmission by beetles, narrow host range)
Bean mosaic 4: *see* Southern bean mosaic	
Bean pod mottle	COMOVIRUS
Bean rugose mosaic	COMOVIRUS
Bean (Southern) mosaic: *see* Southern bean mosaic	
Bean summer death	possibly geminivirus, identical to tobacco yellow dwarf
Bean yellow mosaic	POTYVIRUS, identical to pea mosaic
Bearded iris mosaic	POTYVIRUS (narrow host range, no serological relationship to many others)
Beet cryptic	Cryptic virus

Virus	Classification or properties (Group names in Capitals)
Beet curly top	Probably geminivirus (wide host range, transmitted by lace bug)
Beet leaf curl	PHYTORHABDOVIRUS (RHABDOVIRIDAE), (lace-bug-transmitted)
Beet mild yellowing	LUTEOVIRUS
Beet mosaic	POTYVIRUS
Beet necrotic yellow vein[8]	Possibly tobamovirus. Rigid helical rods (2.6 nm pitch), (20 × 85, 100, 265, 390 nm) consisting of four RNAs of 0.6, 0.7, 1.8, and 2.3×10^6 daltons and a 21×10^3 dalton coat protein with 197 amino acids. RNAs are capped and have 5′ terminal poly(A). Transmitted by the fungus *Polymyxa betae*. Narrow host range. In most regards closely similar to potato mop top and soil-borne wheat mosaic virus, and possibly forming a subgroup with these.
Beet Western leaf roll	probably luteovirus
Beet Western yellows	LUTEOVIRUS
Beet yellows	CLOSTEROVIRUS, subgroup I
Beet yellow net	possibly luteovirus
Beet yellow stunt	CLOSTEROVIRUS, subgroup I (12.5 × 1400 nm, 6% RNA of 4.6×10^6 daltons)
Belladonna mosaic	strain of tobacco rattle virus
Belladonna mottle	TYMOVIRUS, serologically related to Andean potato latent virus
Beta 2	synonym for beat mosaic virus
Beta 4	synonym for beet yellows virus
Bidens mosaic	possibly potyvirus
Bidens mottle	POTYVIRUS
Blackeye cowpea mosaic	POTYVIRUS
Blackgram mottle[9]	unclassified (28 nm isometric particles, 122 S, 1.4×10^6 daltons RNA, 38×10^3 daltons protein, transmitted by beetles, serologically unrelated to all viruses tested)

Virus	Classification or properties (Group names in Capitals)
Black locust tree mosaic	synonym for rabinia mosaic virus
Black raspberry latent	strain of tobacco streak virus
Blueberry leaf mottle	NEPOVIRUS
Blueberry shoestring	possibly sobemovirus (transmitted by aphids, no serological relationship detected)
Bobone disease	PHYTORHABDOVIRUS (RHABDOVIRIDAE), (leaf-hopper-transmitted)
Boletus	possibly potexvirus
Boussignaultia mosaic	POTEXVIRUS
Brassica 1	synonym for turnip mosaic
Brassica 3	synonym for cauliflower mosaic virus
Brassica octahedron	synonym for turnip yellow mosaic virus
Brimjal mosaic	unclassified
Broad bean mottle (BBMV)	BROMOVIRUS
Broad bean necrosis[10]	Possibly tobamovirus (25 × 150 and 250 nm, broad host range, serologically related to several tobamoviruses, transmitted by soil-borne fungi)
Broad bean stain	COMOVIRUS
Broad bean true mosaic	COMOVIRUS, synonym for echtes Ackerbohnenmosaik Virus
Broad bean wilt[11]	possibly comovirus (40 × 10³ dalton protein, not serologically related to other comoviruses, has wider host range and is aphid-transmitted, serologically related or identical with nasturtium ringspot and petunia ringspot)
Broccoli mosaic	synonym for cauliflower mosaic virus
Broccoli necrotic yellows	PHYTORHABDOVIRUS (RHABDOVIRIDAE), subgroup A (aphid-transmitted)
Brome (grass) mosaic (BMV)	type member of BROMOVIRUS group

Virus	Classification or properties (Group names in Capitals)
Bromestem leaf mottle	unclassified (structurally and serologically related to phleum mottle and tobacco necrosis virus)
BROMOVIRUS GROUP[12]	Four viruses and their strains make up this group: *Brome mosaic* virus (BMV), (the type member), broadbean mottle virus (BBMV), cowpea chlorotic mottle virus (CCMV), and melandrium yellow fleck virus. The first 3 differ in their isoelectric points (about 7, 5.6, and 4, respectively) and in their thermal inactivation points (about 77, 93, and 78°C).

The icosahedral particles (26 nm diameter, 85 S) differ slightly in buoyant density (about 1.35 g/cm^3 in CsCl), due to slightly different RNA contents which, in turn, are due to the distribution of the four RNAs among them: The densest contains RNA 1 of 1.1×10^6 daltons, the middle component RNA 3 + 4 of 0.75 and 0.3×10^6, and the lightest RNA 2 of 1.0×10^6 daltons. Of the RNAs, only 1 to 3 are required for infectivity. Of the coat protein of 20×10^3 daltons and known sequence, 180 molecules are arranged in a T = 3 lattice (12 pentamers and 20 hexamers). The particles have a hollow center and swell above pH 6.5. The RNAs of BMV are capped (*see* alfalfa mosaic virus); their 3′ terminus can bind tyrosine under conditions of tRNA amino acid acylation. These RNAs have been sequenced. The other bromoviruses have been studied less. BMV and CCMV show a slight serological relationship; a pseudorecombinant of BMV RNA 1 and 2 with CCMV RNA 3 has shown some infectivity. BBMV is not related to these. None of these viruses are serologically related to the cucumovirus which they resemble in most respects.

The replication and translation strategies of the bromoviruses are the same as of the alfalfa mosaic virus and the ilarvi-

Virus	Classification or properties (Group names in Capitals)
BROMOVIRUS GROUP[12] (continued)	ruses. The bromoviruses have narrow host ranges. They appear to be transmitted mostly mechanically, by sap, and occasionally by beetles.
Bryonia mottle	possibly potyvirus
Burdock yellows	CLOSTEROVIRUS, subgroup I
Cabbage A	synonym for turnip mosaic virus
Cabbage B	synonym for cauliflower mosaic virus
Cabbage black ringspot	synonym for turnip mosaic virus
Cabbage mosaic	synonym for cauliflower mosaic
Cacao mottle leaf	unclassified (similar to cacao swollen shoot)
Cacao necrosis	probably nepovirus
Cacao swollen shoot	probably phytorhabdovirus (Rhabdoviridae), (bacilliform rods 28 × 125 nm, 218 S, transmitted by mealybugs, strain S said to be identical to cacao necrosis virus)
Cacao yellow mosaic	TYMOVIRUS
Cactus 2	CARLAVIRUS
Cactus X	POTEXVIRUS
Cadang-cadang	VIROID of palms
Callistephus chinensis chlorosis	possibly phytorhabdovirus (Rhabdoviridae)
Canavalia maritima mosaic	possibly potyvirus
Cantaloupe mosaic	synonym for watermelon mosaic
Caper vein banding	possibly carlavirus
CAPILLOVIRUS GROUP	Proposed group of filamentous viruses to include apple stem grooving and potato virus T.
CARLAVIRUS GROUP[3]	Slightly flexuous filaments or bent rods of 600–700 nm length and 12–15 nm diameter (about 160 S, buoyant density about 1.3 g/cm^3). The protein of 32 to 36 × 10^3 daltons is arranged in a helix of 3.4 nm pitch. The single RNA molecules of var-

Virus	Classification or properties (Group names in Capitals)
CARLAVIRUS GROUP[3] (continued)	ious viruses have molecular weights of $2.3–3.0 \times 10^5$ (6% of virion).
	Carnation latent is the type member and 23 other viruses are identified as members and a similar number as possible members or strains. Not all are serologically related. Most of them have narrow host ranges; but carlaviruses, in general, infect alfalfa, cactus, carnations, chrysanthemum, elderberries, hops, lilac, lilies, peas, potatoes, clover, etc. Often infection is not evident (latent). Transmission is usually by aphids in nonpersistent manner.
Carna 5: *see* Satellite RNAs	
Carnation bacilliform	possibly phytorhabdovirus (Rhabdoviridae)
Carnation cryptic	cryptic viruses
Carnation etched ring	CAULIMOVIRUS (double-stranded DNA)
Carnation Italian ringspot	TOMBUSVIRUS
Carnation latent	type member of CARLAVIRUS group
Carnation mottle[4]	unclassified (structurally and biologically similar to tombusvirus, no serological relationship detected)
Carnation necrotic fleck	CLOSTEROVIRUS, subgroup I
Carnation ringspot	type member of DIANTHOVIRUS group
Carnation streak	possibly closterovirus subgroup B
Carnation vein mottle	POTYVIRUS
Carnation yellow flex	probably closterovirus subgroup I, transmitted by aphids
Carrot latent	Probably phytorhabdovirus (Rhabdoviridae), aphid-transmitted
Carrot mosaic	possibly potyvirus
Carrot motley dwarf	synonym for carrot mottle virus
Carrot mottle[15]	Defective LUTEOVIRUS (helper: carrot red leaf virus, forms enveloped hetero-disperse (av-

Virus	Classification or properties (Group names in Capitals)
Carrot mottle[15] (*continued*)	erage 270 S, 52 nm diameter, buoyant density in CsCl 1.15 g/cm^3) particles; RNA 1.55×10^6 daltons, no terminal poly(A), transmitted mechanically and in persistent manner by aphids. *See* lettuce speckles for similar relation to beet Western yellows.
Carrot red leaf	LUTEOVIRUS
Carrot thin leaf	POTYVIRUS
Carrot yellow leaf	CLOSTEROVIRUS, subgroup I
Cassava common mosaic	POTEXVIRUS
Cassava latent	GEMINIVIRUS (single-stranded DNA)
Cassava symptomless	possibly phytorhabdovirus (Rhabdoviridae)
Cassava vein mosaic	possibly caulimovirus
Cassia mild mosaic	possibly carlavirus
Cauliflower mosaic	type member of CAULIMOVIRUS group
CAULIMOVIRUS GROUP[16]	A small group of double-stranded (ds) DNA-containing viruses; type member is the *cauli*flower *mo*saic virus, and others, serologically related, are carnation etched ring, dahlia mosaic, strawberry vein banding virus and probably others. Spherical particles of about 50 nm diameter (208 S, density 1.37 g/cm^3 in CsCl) (Figure 19), containing 17% of a single molecule of largely circular and double-stranded DNA of about 5×10^6 daltons. The DNA has interruptions, one in one strand and two in the other. It has been sequenced. The multiple proteins observed may, in part, be artifacts due to degradation and aggregation, the coat protein being about 50×10^3 daltons. These viruses have limited host ranges. Cauliflower mosaic virus is transmitted by aphids in nonpersistent manner. The severity of symptoms varies greatly for different strains. The members of the group are serologically related.

Virus	Classification or properties (Group names in Capitals)
Celery mosaic	POTYVIRUS
Celery yellow mosaic	possibly potyvirus
Celery yellow spot	possibly luteovirus
Centrosema mosaic	possibly potexvirus, aphid-transmitted
Cereal chlorotic mottle	PHYTORHABDOVIRUS (RHABDOVIRIDAE)
Cereal striate	PHYTORHABDOVIRUS (RHABDOVIRIDAE)
Cereal tillering disease	FIJIVIRUS (REOVIRIDAE), serologically related to maize rough dwarf virus
Cereal yellow dwarf	synonym for barley yellow dwarf virus
Chara australis (algae)	possibly tobamovirus (*see* chara corallina)
Chara corallina (algae)[17]	probably tobamovirus (18 × 532 nm rods, helical pitch 2.7 nm, 230 S, 5% RNA of 3.6 × 10 daltons, 17.5 × 10^3 dalton protein, distant serological relation to all, most to CV4)
Chenopodium mosaic	synonym for sowbane mosaic virus
Cherry chloratic ringspot	synonym for prune dwarf virus
Cherry leaf roll	NEPOVIRUS (RNA 1, 2: 2.4, 2.0 × 10^6 daltons)
Cherry rasp leaf	possibly nepovirus
Cherry rugose mosaic	ILARVIRUS
Cherry (sour) necrotic ringspot	synonym for prunus necrotic ringspot
Chicory blotch	possibly carlavirus
Chicory yellow mottle	NEPOVIRUS
Chinese rape mosaic	probably tobamovirus
Chloris striate mosaic	GEMINIVIRUS (single-stranded DNA)
Chlorotic leaf spot	CLOSTEROVIRUS
Chondrilea juncea stunting	possibly phytorhabdovirus (Rhabdoviridae)
Chrysanthemum aspermy	CUCUMOVIRUS
Chrysanthemum B	CARLAVIRUS

Virus	Classification or properties (Group names in Capitals)
Chrysanthemum chlorotic leafspot	possibly closterovirus (no serological relationship to others, not aphid- nor seed-transmitted)
Chrysanthemum chlorotic mottle	VIROID
Chrysanthemum frutescens	probably phytorhabdovirus (Rhabdoviridae)
Chrysanthemum mosaic	synonym for chrysanthemum aspermy virus
Chrysanthemum stunt[18]	VIROID (359 nucleotides, 69% of sequence identical to that of potato spindle tuber viroid)
Chrysanthemum vein chlorosis	possibly phytorhabdovirus (Rhabdoviridae)
Citrus exocortis[19]	VIROID (371 nucleotides, sequences differ for different isolates)
Citrus leaf rugose	ILARVIRUS
Citrus leprosis	possibly nonenveloped phytorhabdovirus (Rhabdoviridae)
Citrus stunt-tatter leaf	unclassified filamentous particles
Citrus tristeza	CLOSTEROVIRUS, subgroup I
Citrus variegation	ILARVIRUS
Clitoria yellow vein	TYMOVIRUS
CLOSTEROVIRUS GROUP[20]	A poorly defined group of rather labile viruses. Thin (12 nm diameter), very flexuous filaments (helical pitch 3.7 nm). Subgroups of different lengths, I: 1250–1800 nm, and II: 600–750 nm (about 96 S); buoyant density in CsCl about 1.32 g/cm³) have been recognized. The corresponding RNAs (5%) have molecular weights from 6.5–2.5×10^6, respectively. A serological relationship has been detected only between beet yellows, wheat yellow leaf, and carnation necrotic flex virus. The closteroviruses of subgroup I and II have coat proteins of about 24 and 27×10^3 daltons,

Virus	Classification or properties (Group names in Capitals)
CLOSTEROVIRUS GROUP[20] (continued)	respectively; several lack tryptophan. Closteroviruses (Greek Kloster = thread) of subgroup I are transmitted by aphids, some by a single aphid species, but not generally those of subgroup II. These also lack the characteristic vesicular inclusion bodies of subgroup I. Classification as three groups (A, <800 nm; B, 1250–1450 nm; C, 1600–2000 nm) has also been proposed.
Clover big vein	synonym for wound tumor virus
Clover (Croatina)	possibly potyvirus
Clover enation	possibly phytorhabdovirus (Rhabdoviridae)
Clover mild mosaic	unclassified, 28 nm particles, aphid-transmitted
Clover primary leaf necrosis	DIANTHOVIRUS
Clover wound tumor: see wound tumor virus	
Clover yellow mosaic	POTEXVIRUS
Clover yellow vein	POTYVIRUS
Clover yellows	CLOSTEROVIRUS, subgroup I
Cocksfoot mild mosaic[21]	unclassified (28 nm icosahedral particles (105 S), 24% RNA, 25 × 10³ dalton protein, serologically closely related to phleum mottle virus)
Cocksfoot mottle	possibly sobemovirus (no serological relationship detected, except to cynosorus mottle)
Cocksfoot streak	POTYVIRUS
Cocoa necrosis	NEPOVIRUS
Cocoa mottle leaf	probably phytorhabdovirus (Rhabdoviridae)
Cocoa swollen shoot	probably phytorhabdovirus (Rhabdoviridae)
Cocoa yellow mosaic	TYMOVIRUS
Coconut Cadang-Cadang: see Cadang-Cadang	
Coffee ringspot	probably phytorhabdovirus (Rhabdoviridae)

Virus	Classification or properties (Group names in Capitals)
Cole latent	possibly carlavirus
Colocasia alomea disease	related to cacao swollen shoot virus
Colocasia bobone disease	probably phytorhabdovirus (Rhabdoviridae), (leaf hopper-transmitted).
Columbi datura	POTYVIRUS
Columnea	VIROID
Commelina mosaic	POTYVIRUS
COMOVIRUS GROUP[22]	Two types of nucleoprotein particles of 28 nm diameter (118 S and 98 S, density in CsCl 1.41 and 1.44 g/cm^3) containing 34 and 25% of RNA (2.4 and 1.4 $\times$ 10^6 daltons) frequency also nucleic acid-free 58 S "top component" particles. In all particles, there are 60 molecules each of two different coat proteins of 22 and 42 $\times$ 10^3 daltons. Both heavy particles, or their RNAs are required for infectivity. Pseudorecombinants of the RNAs of different members of the group are infectious. *Cowpea mosaic* virus is the type member. Its RNA and probably others, have a covalently-linked 5' terminal protein, the removal of which does not cause loss of infectivity. Several comoviruses, if not all, have 3' terminal poly A. At least 12 members are known, namely, bean curly dwarf, bean pod mottle, bean rugose mosaic, broad bean stain, broad bean true mosaic, cowpea severe mosaic, cowpea mosaic, pea green mottle, quail pea mosaic, red clover mottle, all *Leguminosae* as principal host; radish mosaic, squash mosaic, and Andean potato mottle virus of *Cruciferae*, *Cucurbitaceae*, and *Solanaceae*, respectively. All show at least one serological interrelationship. Cowpea mosaic virus shows serological relationships with nine others. The host range of most comoviruses is narrow. Relationship to Picornaviridae has been demonstrated.

Virus	Classification or properties (Group names in Capitals)
Cotton anthocyanosis	LUTEOVIRUS
Cow parsnip mosaic	probably phytorhabdovirus (Rhabdoviridae)
Cowpea aphid-borne mosaic	POTYVIRUS
Cowpea chlorotic mottle (CCMV)[23]	BROMOVIRUS
Cowpea mild mottle	CARLAVIRUS (no known vectors)
Cowpea (yellow) mosaic (CPMV)[24]	type member of COMOVIRUS group
Cowpea mottle[25]	unclassified (isometric 30 nm diameter particles, RNA of 1.4×10^6, protein of 45×10^3 daltons, wide host range, transmitted by beetles, serologically not related to many isometric viruses)
Cowpea ringspot	probably cucumovirus
Cowpea severe mosaic[26]	COMOVIRUS
Cowpea strain of tobacco mosaic: *see* sunnhemp mosaic virus	
Crimson clover latent	NEPOVIRUS
Crinum	possibly potyvirus
Croatian clover: *see* clover (Croatian) virus	
CRYPTIC VIRUSES[27]	Noninfectious symptomless viruses, many transmittable apparently only through seeds, occurring as particles of about 30 nm diameter at very low concentrations. The nucleic acid of only one (carnation cryptic virus) has been characterized as three double-stranded RNA segments, totaling 4×10^6 daltons. Others are beet cryptic, vicia cryptic, poinsettia cryptic, and rye grass spherical virus.
Cucumber 1	synonym for cucumber mosaic virus
Cucumber 3 and 4 (CV3, CV4)	TOBAMOVIRUSES (distantly related to tobacco mosaic virus)
Cucumber fruit streak	unclassified 30 nm diameter particles (132 S) containing RNA of 1.45×10^6 daltons

Virus	Classification or properties (Group names in Capitals)
Cucumber green mottle mosaic (CGMMV)	TOBAMOVIRUS (serologically distantly related to tobacco mosaic virus)
Cucumber mosaic (CMV)[28]	type member of CUCUMOVIRUS group
Cucumber necrosis[29]	unclassified (isometric 31 nm diameter particles, 133 S, containing RNA of 1.4×10^6 daltons, soil-borne, fungus-transmitted; serologically not related; possibly necrovirus group, though small host range)
Cucumber pale fruit	VIROID
Cucumber soilborne	(unclassified 31 nm particles, 120 S protein of 4.14×10^3 and RNA of 1.5×10^6 daltons)
Cucumber vein yellowing[30]	unclassified (15×750 nm particles consisting of 39×10^3 dalton protein and double-stranded DNA, transmitted by white flies)
Cucumber yellow mosaic	strain of cucumber mosaic virus
Cucumber, yellows	unclassified (12×1000 nm filaments, white fly transmitted, labile)
CUCUMOVIRUS GROUP[31]	There are four known members of this group: Cucumber mosaic virus (CMV), the type member, peanut stunt, robinia mosaic, and tomato aspermy virus. The virus particles of 99 S are icosahedral with a 28 nm diameter (density in CsCl 1.37 g/cm^3). They contain either RNA 1, 2, or 3 + 4 of about 1.3, 1.1, and $0.8 + 0.3 \times 10^6$ daltons and a protein of 24×10^3 daltons. The particles sediment together. All three types of particles, or the three largest RNAs, are required for infectivity. The RNAs are 5′ capped (see alfalfa mosaic virus) and their 3′ ends accept tyrosine under the conditions of tRNA aminoacylation. The cucumoviruses have a wide host range, they occur worldwide, and are transmitted by aphids in nonpersistent manner. Serological relationship between members of the

Virus	Classification or properties (Group names in Capitals)
CUCUMOVIRUS GROUP[31] (continued)	group and various strains range from near to very far. The resemblance of the chemical properties of the cucumo and bromoviruses is great but, biologically, they are clearly different. Some resemblance of the cucumoviruses to alfalfa mosaic virus in chemical as well as biological properties is also evident.
Curly top	GEMINIVIRUS (single-stranded DNA).
CV 3, 4: see cucumber virus 3, 4	
Cymbidium mosaic	POTEXVIRUS
Cymbidium ringspot	possibly tombusvirus (structurally similar, but no serological relationship to 18 tombus and many other isometric viruses)
Cynara	probably phytorhabdovirus (Rhabdoviridae)
Cynodon mosaic	CARLAVIRUS
Cynosorus mottle[32]	unclassified (serologically related to cocksfoot mottle virus, phleum mottle virus)
Dahlemense	strain of tobacco mosaic virus (protein sequenced, not closely related to U1)
Dahlia mosaic	CAULIMOVIRUS (double-stranded DNA)
Daikon mosaic	synonym for turnip mosaic
Danis plum line pattern	ILARVIRUS (identical to apple mosaic)
Daphne X	possibly potexvirus (no known vector nor serological relationship)
Daphne Y	possibly potyvirus
Dasheen mosaic	POTYVIRUS
Datura 437	possibly potyvirus
Datura mosaic	possibly potyvirus
Datura shoestring	POTYVIRUS (serological relationship to pepper veinal mottle virus)
Dendrobium leaf streak	possibly nonenveloped phytorhabdovirus (Rhabdoviridae)
Dendrobium mosaic	possibly potyvirus

Virus	Classification or properties (Group names in Capitals)
Desmodium	COMOVIRUS
Desmodium mosaic	possibly potyvirus
Desmodium yellow mottle	TYMOVIRUS (no known vector)
DIANTHOVIRUS GROUP[33]	A small group of polyhedral viruses of about 30 nm diameter, 7×10^6 particle weight (135 S) density in CsCl about 1.37 g/cm³, containing two RNAs of 1.5 and 0.5×10^6 daltons and a coat protein of 40×10^3 daltons. The larger RNA carries the coat protein gene. Carnation (Latin name: *dianthus*) ringspot is the type member, but serologically unrelated to red clover and sweet clover necrotic mosaic viruses, the only other known members of the group. These viruses have wide host ranges; they are transmitted through the soil. Pseudorecombinants between various dianthoviruses are infectious.
Digitaria striate	probably phytorhabdovirus (Rhabdoviridae), (hopper-transmitted)
Dioscorea green-banding	possibly potyvirus
Dioscorea latent	possibly potexvirus
Dioscorea trifida	possibly potyvirus
Dock mottling mosaic	possibly potyvirus
Dolichos enation mosaic	strain of tobacco mosaic virus (legumes main host)
Dulcamara mottle	TYMOVIRUS, possibly strain of belladonna mottle virus
Echtes Ackerbohnenmosaik: *see* broadbean true mosaic (comovirus)	
Echtes Robinienmosaik	synonym for robinia mosaic
Eggplant mild mottle	CARLAVIRUS
Eggplant mosaic	TYMOVIRUS
Eggplant mottled crinkle	TOMBUSVIRUS

Virus	Classification or properties (Group names in Capitals)
Eggplant mottled dwarf	PHYTORHABDOVIRUS (RHABDOVIRIDAE), subgroup B
Elderberry latent	possibly tombusvirus (no serological relation to any isometric virus tested)
Elderberry latent A	CARLAVIRUS
Elm mosaic	possibly identical with cherry leaf roll virus
Elm mottle	ILARVIRUS (wide host range, no known vectors, no serological relationship detected)
Enation pea	synonym for pea enation virus
Endive	probably phytorhabdovirus (Rhabdoviridae)
Erbsenstauche	synonym for red clover vein mosaic virus
Erysimum latent	TYMOVIRUS
Eupatorium yellow vein (or leaf)	probably geminivirus (single-stranded DNA)
Euphorbia mosaic	GEMINIVIRUS (single-stranded DNA)
Euphorbia ringspot	possibly potyvirus
European wheat striate mosiac	possibly rice stripe virus group
F6-A, F6-B, F6-C	MYCOVIRUS (double-stranded RNA)
Fern	possibly potyvirus
Festuca leaf streak	possibly phytorhabdovirus (Rhabdoviridae)
Festuca necrosis	CLOSTEROVIRUS, subgroup I
Figwort mosaic	CAULIMOVIRUS (double-stranded DNA)
Fiji disease	type member of FIJI VIRUS (REOVIRIDAE) (serologically unrelated to others)
FIJI VIRUSES[34]	A genus of the Reoviridae differing from the genus Phytoreoviruses in being slightly smaller (65 nm diameter). Their outer protein shell is less stable, thus revealing the spikes more readily. Further, their genome consists of only ten RNA components (of $1.0–2.9 \times 10^6$ daltons). Each RNA molecule codes for a protein, six of which make up the virion.

Virus	Classification or properties (Group names in Capitals)
Fɪᴊɪ ᴠɪʀᴜsᴇs[34] (*continued*)	The type member is Fiji disease virus, others cereal tillering disease, maize rough dwarf, pangola stunt, rice black streaked dwarf, *Arrhenatherum* blue dwarf, *Lolium* enation, and oat sterile dwarf viruses. These belong to at least three serologically unrelated subgroups.
Filaree red leaf	possibly luteovirus
Finger millet mosaic	possibly phytorhabdovirus (Rhabdoviridae)
Flat apple	synonym for cherry rasp leaf virus
Foxtail mosaic	Pᴏᴛᴇxᴠɪʀᴜs (serologically related to narcissus mosaic and viola mottle virus)
Frangipani mosaic	Tᴏʙᴀᴍᴏᴠɪʀᴜs (physically and serologically similar to tobacco mosaic virus, narrow host range)
Freesia mosaic	possibly potyvirus (aphid-transmitted)
Fuchsia latent	possibly carlavirus
Fungal viruses	*see* mycoviruses
Gaeumannomyces spp.	Mʏᴄᴏᴠɪʀᴜs (double-stranded RNA)
Garlic mosaic	possibly carlavirus
Garlic yellow streak	possibly potyvirus
Gᴇᴍɪɴɪ ᴠɪʀᴜs ɢʀᴏᴜᴘ[35]	A group of viruses characterized by small (17 × 33 nm) particles occurring usually as Siamese twins (*gemini*), (70 S) containing each a circular single-stranded plus-strand DNA of about 0.8×10^6 daltons, differing in sequence, and both needed for infection. The coat protein is about 31×10^3 daltons. Generally narrow host ranges, transmission by leaf hoppers or white flies in persistent manner and, not easily, by mechanical means. Members are maize streak, bean golden mosaic, cassava latent, *Chloris* striate mosaic and tomato golden mosaic virus. No serological relationships have been demonstrated.
Gerbera symptomless	possibly phytorhabdovirus (Rhabdoviridae)

Virus	Classification or properties (Group names in Capitals)
Gloriosa stripe mosaic	possibly potyvirus
Glycine mosaic	COMOVIRUS
Glycine mottle	possibly tombusvirus
Golden elderberry	synonym for cherry leaf roll virus
Gombo mosaic	TYMOVIRUS
Gomphrena	possibly phytorhabdovirus (Rhabdoviridae)
Grape decline	synonym for peach rosette mosaic virus
Grapevine ajinashika	possibly luteovirus
Grapevine Bulgarian latent	possibly nepovirus (no serological relationship to other nepoviruses, not transmitted by nematodes)
Grapevine chrome mosaic	possibly nepovirus (no serological relationship to other nepoviruses, except cocoa necrosis virus, soil-transmitted)
Grapevine degeneration	synonym for peach rosette mosaic
Grapevine fanleaf	NEPOVIRUS (serologically related to arabis mosaic virus)
Grapevine leaf roll	possibly identical with grapevine stem pitting associated virus
Grapevine stem pitting associated[36]	CLOSTEROVIRUS, subgroup II
Grapevine yellow vein	similar or identical with tomato ringspot virus
Grass mosaic	synonym for sugarcane mosaic
Grassy stunt	see rice grassy stunt
Green aucuba: *see* Aucuba virus	
Groundnut crinkle	possibly carlavirus
Groundnut eyespot	possibly potyvirus
Groundnut rosette (assistor)	possibly luteovirus (it appears that this virus is generally associated with another unidentified symptomless virus, the assistor, and that aphid transmission requires the presence of both)

Virus	Classification or properties (Group names in Capitals)
GTAMV	strain of tobacco mosaic virus, not closely related
Guar symptomless	possibly potyvirus
Guinea grass mosaic	POTYVIRUS
Gynura latent	possibly strain of Chrysanthemum B
Helenium S	CARLAVIRUS
Helenium Y	possibly potyvirus
Helminthosporium spp.	MYCOVIRUS (double-stranded RNA)
Henbane mosaic	POTYVIRUS
Heracleum latent	possibly closterovirus, subgroup II
Heracleum 6	CLOSTEROVIRUS, subgroup I
Hibiscus chlorotic ringspot[37]	unclassified (28 nm diameter particles, RNA of 1.6×10^6 daltons, serologically not related to 45 other isometric viruses, vector unknown)
Hibiscus (latent) ringspot	NEPOVIRUS
Hippeastrum latent	possibly potexvirus
Hippeastrum mosaic	POTYVIRUS
Hoja blanca	unclassified (filamentous plant hopper-transmitted virus)
Holcus lanatus yellowing	possibly phytorhabdovirus (Rhabdoviridae)
Holcus streak	possibly potyvirus
Holcus transitory mottle	unclassified, similar to brome stem leaf mottle virus
Holmes ribgrass (HR)	TOBAMOVIRUS, distantly related to tobacco mosaic virus (156 amino acids in sequenced protein)
Honeysuckle latent	CARLAVIRUS
Hop A	ILARVIRUS
Hop B	ILARVIRUS (identical to prunus necrotic ringspot)

Virus	Classification or properties (Group names in Capitals)
Hop C	ILARVIRUS
Hop latent	CARLAVIRUS
Hop mosaic	CARLAVIRUS
Hop stunt	VIROID
HORDEI VIRUS GROUP[38]	Rigid rods of varying lengths (about 22 × 100–150 nm) (178–200 S) containing 2–4 molecules of RNA (4%, 1.0–1.5 × 10^6 daltons each), varying for different strains and viruses and not all nor the same ones necessary for infectivity. Barley (Latin: *hordeum*) stripe mosaic, the type member, carries varying amounts of poly(A) near but not at the 3′ end of most RNAs, the 3′ end being able to bind tyrosine under tRNA charging conditions; the 5′ ends are capped (*see* alfalfa mosaic virus). The single coat protein has a molecular weight of 21 × 10^3 and is glycosylated. Other members are *lychnis* ringspot and poa semilatent virus. Their host ranges are narrow, and transmission is through seed. Serologically, they are distantly related.
Hordeum mosaic	possibly potyvirus
Hordeum nanescens	synonym for barley yellow dwarf
Horseradish mosaic	synonym for turnip mosaic
HR: *see* Holmes ribgrass	
HVCV-1 and 2	ALGAL VIRUSES (from Chlorella, symbiotic with Hydra)
HvV-A	MYCOVIRUS (double-stranded RNA)
Hyacinth mosaic	possibly potyvirus
Hydrangea ringspot	POTEXVIRUS
Hydra viridis	ALGAL VIRUS
Hyoscyamus I, III	synonym for henbane mosaic
Hypochoeris mosaic	possible tobamovirus (fragile rods 22 × 130 and 22 × 230 nm, serologically unrelated to other tobamoviruses, coat protein of 25 × 10^3 daltons).

Virus	Classification or properties (Group names in Capitals)
ILARVIRUS GROUP[39]	Particles vary from isometric to bacilliform with 26–35 nm diameters, 80–120 S. Their density of 1.36 g/cm^3 in CsCl differs from that of the otherwise similar alfalfa mosaic virus (1.28 g/cm^3). They contain four molecules of RNA of about 1.1, 0.9, 0.7, and 0.3 × 10^6 daltons. The smallest, the coat protein gene, is not required for infectivity if coat protein is present, the same situation as for alfalfa mosaic virus. The name is derived from *i*sometric *l*abile ringspot virus. The coat protein is 25 × 10^3 daltons. The type member, tobacco streak virus, has a wide host range, and is transmitted through seeds and pollen, but not by any insects. Others are apple mosaic, citrus leaf rugose and variegation, elm mottle, lilac ring mottle, North American plum line pattern, prune dwarf, *prunus* necrotic ringspot, spinach latent, and Tulare apple mosaic viruses. Subgroups of the ilarviruses are serologically interrelated.
Indonesian soybean dwarf	LUTEOVIRUS
Iris fulva mosaic	possibly potyvirus
Iris germanica leaf stripe	possibly phytorhabdovirus (Rhabdoviridae)
Iris mild mosaic	POTYVIRUS
Iris severe mosaic	POTYVIRUS
Isachne mosaic	possibly potyvirus
Ivy vein clearing	possibly phytorhabdovirus (Rhabdoviridae)
J14D1	strain of tobacco mosaic virus, closely related, lethal to some tobacco strains
Jonquil mild mosaic	possibly potyvirus (identical to narcissus late season yellows)
Kale	strain of radish mosaic virus
Kartoffel-K	synonym for potato virus M
Kennedya Y	possibly potyvirus

Virus	Classification or properties (Group names in Capitals)
Kennedya yellow mosaic	TYMOVIRUS
Laburnum yellow vein	possibly phytorhabdovirus (Rhabdoviridae)
Lactuca	synonym for lettuce mosaic virus
Laelia red leaf spot	possibly phytorhabdovirus (Rhabdoviridae)
Launea arborescent stunt	possibly phytorhabdovirus (Rhabdoviridae)
Leek yellow strip	POTYVIRUS
Legume yellows	LUTEOVIRUS
Lemon-scented thyme leaf chlorosis	possibly phytorhabdovirus (Rhabdoviridae)
Lettuce mosaic	POTYVIRUS
Lettuce necrotic yellow	PHYTORHABDOVIRUS (RHABDOVIRIDAE), type member of subgroup A (aphid-transmitted)
Lettuce speckless mottle[40]	Unclassified (unstable infectivity probably a defective, coat gene-lacking virus, depending on beet Western yellows as a helper virus for aphid transmission). Identification of toga virus-like particles is of doubtful significance. The infective RNA is 1.4×10^6 daltons. *See* carrot mottle virus for similar situation.
Lilac chlorotic leafspot	probably closterovirus, subgroup I (no serological relationship to others detected)
Lilac mottle	CARLAVIRUS
Lilac ring mottle	ILARVIRUS
Lily	possibly potexvirus
Lily curly strip	synonym for lily symptomless
Lily symptomless	CARLAVIRUS
Lily X	POTEXVIRUS
Line	unclassified 30-nm-diameter maize virus
Lolium	possibly phytorhabdovirus (Rhabdoviridae)

Virus	Classification or properties (Group names in Capitals)
Lolium enation	FIJI VIRUS (REOVIRIDAE), (identical to *arrhenatherum* blue dwarf)
Lonicera latent	CARLAVIRUS
Lotus stem necrosis	possibly phytorhabdovirus (Rhabdoviridae)
Lucerne Australian latent	NEPOVIRUS
Lucerne enation	possibly phytorhabdovirus (Rhabdoviridae), (aphid-transmitted)
Lucerne mosaic	synonym for alfalfa mosaic virus
Lucerne transient streak	possibly sobemovirus, 1.4×10^6 dalton RNA + 0.12×10^6 dalton circular satellite RNA
LUTEOVIRUS GROUP[41]	Isometric virions of about 28 nm diameter (120 S) consisting of RNA of 2×10^6 daltons, and a coat protein of 24×10^3 daltons. The RNA has a 5′ terminal protein that is not essential for infectivity, and no 3′ terminal poly(A). A rather large group of viruses, some with narrow and others with wide host ranges. All are transmitted by aphids in persistent manner, some with great species specificity, but probably without replication of the virus in the vector. They are located in the phloem tissue. Mechanical transmission is not possible except of protoplasts. Most luteoviruses are serologically related. The type member is barley yellow dwarf virus, and it and many of the others cause yellowing (thus, luteo from Latin *luteus* = yellow).
Lychnis ringspot	HORDEIVIRUS
Lycopersicum Y	synonym for tomato bushy stunt virus
Macana disease of fique	unclassified (isometric 29 nm diameter particles, transmitted by aphids)
Maclura mosaic[42]	possibly potyvirus (15×674 nm threads, 155 S, density 1.307 in CsCl, 45×10^3 dalton protein, no serological relationship to other filamentous viruses)

Virus	Classification or properties (Group names in Capitals)
Maclura ringspot	identical with cucumber mosaic virus
Maize chlorotic dwarf[43]	Monotypic, or possibly a group including rice spherical tungrovirus (isometric 30-nm-diameter particles, 183 S, density 1.51 g/cm^3 in CsCl, 36% of 3.2×10^6 daltons RNA; transmitted by leafhoppers in semipersistent manner, no serological relationship to other viruses tested including rice tungrovirus). Major maize pathogen.
Maize dwarf mosaic	POTYVIRUS
Maize Hoja blanca	possibly phytoreovirus (Reoviridae)
Maize line	probably identical with maize chlorotic dwarf virus
Maize mosaic	possibly phytorhabdovirus (Rhabdoviridae), (transmitted by leafhoppers)
MAIZE RAYADO FINO GROUP[44]	type member: 32-nm-diameter particles, 120 S and 54 S for top component lacking RNA; RNA 2.2×10^6 and two proteins of 22 and 26×10^3 daltons (1 : 3), narrow host range, transmitted by leafhoppers in persistent manner. Other member: rice grassy stunt.
Maize rough dwarf[45]	FIJIVIRUS (Reoviridae), (serologically closely related to rice black-streaked virus)
Maize streak	GEMINIVIRUS (single-stranded DNA)
Maize stripe[46]	rice stripe virus group
Maize wallaby ear	FIJIVIRUS (Reoviridae)
Malva sylvestris	possibly phytorhabdovirus (Rhabdoviridae)
Malva vein clearing	possibly potyvirus (aphid-transmitted)
Malva veinal necrosis	possibly potexvirus
Malva yellows	LUTEOVIRUS
Marigold mottle	possibly potyvirus (transmitted by aphids)

Virus	Classification or properties (Group names in Capitals)
Melandrium yellow fleck[47]	possibly bromovirus [25-nm-isometric particles (88 S) consisting of a 22 × 10³ protein and four RNAs of 1.2, 1.1, 1.0, 0.3 × 10⁶ daltons, all slightly larger than in other bromoviruses. No serological relationship to any virus, wide host range, seed transmission, no vectors]. The virus is very heat-stable.
Melilotus latent	probably phytorhabdovirus (Rhabdoviridae)
Melon mosaic	synonym for watermelon mosaic virus
Melon necrotic spot	possibly necrovirus group
Melon variegation	possibly phytorhabdovirus (Rhabdoviridae)
Micromonas pusilla (MPV)	algal iridovirus (Iridoviridae)
Milk-vetch dwarf	possibly luteovirus
Millet red leaf	possibly luteovirus
Mirabilis mosaic	CAULIMOVIRUS (double-stranded DNA)
Molinia streak	unclassified (related to *panicum* mosaic virus)
Mulberry latent	CARLAVIRUS
Mulberry ringspot	NEPOVIRUS (no serological relationship to other nepoviruses)
Mungbean mosaic	possibly potyvirus (aphid-borne)
Mungbean mottle	possibly potyvirus (aphid-borne)
Mungbean yellow mosaic	GEMINIVIRUS (single-stranded DNA)
Muscat melon mosaic	synonym for watermelon mosaic virus
Mushroom 4[48] (*Agaricus bisporus*)	MYCOVIRUS (double-stranded RNA)
Mushroom viruses: *see* mycoviruses	
Muskmelon vein necrosis	CARLAVIRUS
Mycogone perniciosa	MYCOVIRUS

Virus	Classification or properties (Group names in Capitals)
MYCOVIRUS GROUPS[49]	Typical mycoviruses are isometric particles of 30–48 nm diameter (120–179 S). They all contain double-stranded RNA, either one, two, or three molecules and totaling $2–6 \times 10^6$ daltons. They have a single coat protein of $40–120 \times 10^3$ daltons. Provisionary classification as three families is as follows: (1) The mycoviruses containing one genomic molecule of RNA $(3.0–5.7 \times 10^6$ daltons, 160–280 S) may be represented by two genera, the *Saccharomyces cerevisiae* and the *Helminthosporium maydis* virus group. The type species of the former is ScVl (about 40 nm diameter, 168 S) with RNA of 3.6×10^6 and a protein of 80×10^3 daltons. Other members are ScV2 and *Ustilago maydis* virus, probable member *Mycogone perniciosa* virus, and possible members *Aspergillus foetidas* and *niger* viruses S (Af-S, An-S), *Gaeumannomyces graminis* viruses 3bla-A and F6-A, *Helminthosporium victoriae* virus (HvV-A), and *Thielaviopsis basicola* viruses. The *Helminthosporium maydis* virus (48 nm diameter, 283 S) has an RNA of 5.7×10^6, a protein of 121×10^3 daltons. (2) The mycoviruses with two genomic RNAs may belong to three genera: *Penicillium stoloniferum* PsV-S, and *G. graminis* virus groups I and II. PcV-S consists of separate particles of 30–34 nm diameter containing 0.94 and 1.1×10^6 daltons RNA; the coat protein is 42×10^3 and a minor protein of 56×10^3 may be the RNA polymerase. Other members, serologically related, occur in *Aspergillus ochraceous* and *Diplocarpon rosae.* Both genera of *Gaeumannomyces graminis* viruses also have divided and separately encapsidated RNAs, larger in group I, and yet larger in group II than in PcV-S (up to 1.6×10^6 daltons). The virions (110 S and 135 S) are 35 nm in diameter, the

Virus	Classification or properties (Group names in Capitals)
MYCOVIRUS GROUPS[49] (continued)	coat proteins about 57 and 70 $\times$ 10^3 daltons. Members of group I of the *G. graminis* viruses are 01916-A, 38-4-A, 01-1-4-A, Og A-B, 3bln-C, F6-C, probably 4519-A, and viruses of *Phialophora* spp. To group II belong TI-A, F6-B, Og A-A, 3bla-B. Some take-all fungal viruses (01916-A and 38-4-A) may carry satellite double-stranded RNAs, besides the genomic ones of 1.27 and 1.19 $\times$ 10^6 daltons coding for 62 $\times$ 10^3 and 55 $\times$ 10^3 proteins (the latter the capsid protein). Particles lacking RNA or carrying single-stranded (m)RNA have also frequently been noted. Mushroom virus 4 of *Agaricus bisporus* and other edible mushrooms may belong to this family but have not been sufficiently well-characterized for classification. (3) Mycoviruses with three genomic RNAs occur mostly in *Penicillium* spp. (*chrysogenum*, *brevicompactum*, and *cyaneo-fulvum*), and possibly in *Helminthosporium victoriae* (HcV-B). Their RNAs of about 2 $\times$ 10^6 daltons are again separately encapsidated in 35–40 nm virions of 150 S. The coat proteins of 115 $\times$ 10^3 daltons are serologically interrelated.
Myrobalan latent ringspot[50]	NEPOVIRUS (frequently carries one or more copies of satellite RNAs), (RNA 1, 2, 3: 2.4, 1.9, 0.45 $\times$ 10^6 daltons)
Nandina mosaic	possibly potexvirus
Narcissus degeneration	POTYVIRUS
Narcissus late season yellows	possibly potyvirus (aphid-transmitted, identical to jonquil mild mosaic)
Narcissus latent	CARLAVIRUS
Narcissus (mild) mosaic	POTEXVIRUS
Narcissus tip necrosis[51]	unclassified (isometric 30-nm-particles (123 S) consisting of 42 (and 39?) $\times$ 10^3 dalton protein and 1.6 $\times$ 10^6 dalton RNA. No se-

Virus	Classification or properties (Group names in Capitals)
Narcissus tip necrosis[51] (*continued*)	rological relationship to other viruses, no known vector).
Narcissus yellow stripe	possibly potyvirus (aphid-transmitted)
Nasturtium mosaic	possibly carlavirus
Nasturtium ringspot[11]	unclassified (probably strain of broad bean wilt virus)
Necrotic fleck	probably closterovirus I
NECROVIRUS GROUP	*see* tobacco necrosis virus
Negro coffee mosaic	possibly potexvirus
NEPOVIRUS GROUP[52]	Icosahedral rather stable particles of 28 nm diameter of about 125, 107, and 52 S containing, respectively, 44% of RNA 1, 34% of RNA 2, or no RNA (buoyant densities in CsCl about 1.52, 1.46, and 1.28 g/cm^3). Satellite RNA, often multiple and of 0.1 × 10^6 daltons, are frequently present in the virions. The molecular weights of the two genomic RNAs are about 2.8 and 1.3–2.4 × 10^6. They have a small protein covalently bound to the 5′ end, the removal of which causes loss of infectivity. Poly(A) is at the 3′ end. The coat protein is about 58 × 10^3 daltons.
	Over 20 members of this group are known, the type member being tobacco ringspot virus; subgroups are serologically interrelated, but not with other subgroups. The host ranges of most of these viruses are wide, the infection often being symptomless. Mechanical and seed transmission are facile, but *nematodes* in the soil are the usual vectors which retain the *poly*hedral virus (ne po) for months without its replicating.
Nerine	possibly potyvirus (aphid-borne)
Nerine latent	CARLAVIRUS
Nerine X	POTEXVIRUS
Nicotiana 7	synonym for tobacco etch virus

Virus	Classification or properties (Group names in Capitals)
Nicotiana 12, 13	synonym for tobacco and tomato ringspot viruses
Nicotiana velutina mosaic[53]	possibly tobamovirus (rods of 18 × 100–175 and 290 nm, 21.4 × 10^3 dalton protein and 2.3 × 10^6 dalton RNA, soil-borne, serologically unrelated to all rod-shaped viruses tested)
Nigerian cowpea	possibly tobamovirus (serologically distantly related to tobacco mosaic virus)
North American plum line pattern	ILARVIRUS
Northern cereal mosaic	possibly phytorhabdovirus (Rhabdoviridae)
Nothoscordum mosaic	POTYVIRUS
06–67	strain of tobacco mosaic virus, closely related to U1
Oat blue dwarf[54]	unclassified (29-nm-isometric particles, 119 S, containing RNA of 2.1 × 10^6 daltons. Wide host range, transmitted by leafhoppers; possibly maize rayado fino group.
Oat mosaic	possibly potyvirus (fungus-transmitted, no serological relationship to other viruses detected)
Oat necrotic mottle	possibly potyvirus (distant serological relationship to wheat streak mosaic virus, mite-transmitted)
Oat red leaf	synonym for barley yellow dwarf
Oat sterile dwarf	FIJIVIRUS (REOVIRIDAE), (identical to *arrhenatherum* blue dwarf)
Oat striate	PHYTORHABDOVIRUS (RHABDOVIRIDAE), (leafhopper-transmitted)
Odontoglossum ringspot	TOBAMOVIRUS (157 amino acids, known sequence, serologically close to tobacco mosaic virus)
Og A-A	MYCOVIRUS (double-stranded RNA)
Og A-B	MYCOVIRUS (double-stranded RNA)
Okra mosaic	THYMOVIRUS

Virus	Classification or properties (Group names in Capitals)
Onion yellow leaf	POTYVIRUS
Ononis yellow mosaic	TYMOVIRUS, possibly strain of *scrophularia* mottle
Orchid fleck	possibly nonenveloped phytorhabdovirus (Rhabdoviridae)
Orchid mosaic	synonym for *cymbidium* mosaic virus
Ornithogalum mosaic	possibly potyvirus (aphid-borne)
Oryza	synonym for rice dwarf virus
Palm mosaic	possibly potyvirus (transmitted by aphids)
Pangola stunt	FIJIVIRUS (REOVIRIDAE), (serologically related to maize rough dwarf virus, etc.)
Panicum mosaic[55]	Unclassified (isometric 28 nm particles, 109S containing 28 S RNA and a 29×10^3 dalton protein. Serologically related to phleum mottle and molinia streak virus. The virus is associated with a satellite virus of 42 S (17 nm diameter) containing 14 S RNA and a 15.5×10^3 dalton protein. (No known vector, limited host range).
Papaya (mild) mosaic[56]	POTEXVIRUS
Papaya ringspot	POTYVIRUS
Paramecium bursaria	ALGAL VIRUS
Parsley latent	PHYTORHABDOVIRUS (RHABDOVIRIDAE), (aphid-transmitted)
Parsley (5)	possibly potexvirus
Parsnip (3)	possibly potexvirus
Parsnip mosaic	POTYVIRUS
Parsnip yellow fleck[57]	unclassified (30 nm diameter, (148 S) particles, RNA of 3.7×10^6 daltons, depends on helper (*anthriscus* yellows) for transmission by aphids.)
Paspalum striate mosaic	probably geminivirus (single-stranded DNA)
Passiflora latent	CARLAVIRUS

Virus	Classification or properties (Group names in Capitals)
Passionfruit ringspot	possibly potyvirus
Passionfruit woodiness	POTYVIRUS
Patchouli mottle	possibly phytorhabdovirus (Rhabdoviridae)
Pawpaw mosaic	synonym for papaya (mild) mosaic virus
PBCV-1	ALGAL VIRUSES
Pea	synonym for pea enation mosaic
Pea early browning	TOBRAVIRUS (105 and 215 nm particles)
Pea enation mosaic[58]	Monotypic: two isometric 27-nm-particles (112 and 99 S) consisting of 22×10^3 dalton protein and two RNAs of about 1.4 and 1.15×10^6 daltons that are required for infectivity, no 3' terminal poly(A), but a 3' terminal protein of 17.5×10^3 daltons; aphid and mechanically transmitted, limited host range, serologically unrelated to other viruses.
Pea leaf roll	LUTEOVIRUS
Pea mosaic	POTYVIRUS (possibly a strain of bean yellow mosaic virus)
Pea necrosis	POTYVIRUS (possibly identical to clover yellow vein virus)
Pea seed-borne mosaic	POTYVIRUS
Pea streak	CARLAVIRUS
Pea symptomless	strain of red clover mottle virus
Peach ringspot	probably identical with *prunus* ringspot virus
Peach rosette mosaic	NEPOVIRUS
Peach stunt	probably identical with prune dwarf virus
Peach yellow bud	strain of tomato ringspot virus
Peanut clamp	possibly identical to peanut clump
Peanut clump[59]	Unclassified (resembles hordei more than tobamo viruses, rigid rods of 21×45 and 190 nm consisting of 2.1 and 1.7×10^6

Virus	Classification or properties (Group names in Capitals)
Peanut clump[59] (continued)	RNA and 23 × 10³ dalton protein. Wide host range, transmitted by fungus *polymyxa*, seed and mechanically; serologically not related to other viruses).
Peanut mottle	POTYVIRUS
Peanut stunt	CUCUMOVIRUS
Peanut yellow mottle	TYMOVIRUS
Pelargonium flower break[60]	unclassified (structurally and biologically similar to tombusvirus, no serological relationships detected, limited host range)
Pelargonium leaf curl	TOMBUSVIRUS
Pelargonium line pattern	unclassified (stable 30-nm-diameter particles, one protein, and one RNA, no serological relationship to 45 isometric viruses tested)
Pelargonium vein clearing	probably phytorhabdovirus (Rhabdoviridae)
Pelargonium zonate spot[61]	unclassified (three components of 25–35 nm diameter, containing 18% RNA of 0.95 and 1.25 × 10⁶ daltons and a protein of 25 × 10³ daltons, differentiating it from ilarviruses).
Penicillium spp.	MYCOVIRUS (double-stranded RNA)
Pepino latent	CARLAVIRUS
Pepino mosaic	POTEXVIRUS
Pepper mottle	POTYVIRUS
Pepper ringspot	strain of tobacco rattle virus
Pepper severe mosaic	POTYVIRUS
Pepper veinal mottle	POTYVIRUS
Peru tomato	POTYVIRUS
Petunia asteroid mosaic	TOMBUSVIRUS
Petunia ringspot	unclassified, possibly strain of broad bean wilt virus

Virus	Classification or properties (Group names in Capitals)
Petunia vein clearing	possibly caulimovirus (double-stranded DNA)
Phalaenopsis chlorotic spot	possibly nonenveloped phytorhabdovirus (Rhabdoviridae)
Phaseolus	synonym for bean common mosaic virus
Phialophora spp.	MYCOVIRUS (double-stranded RNA)
Phleum mottle[62,21]	unclassified (structurally and serologically closely related to panicum and cocksfoot mild mosaic and bromestem leaf mottle virus
Physalis mild chlorosis	possibly luteovirus
Physalis mosaic	TYMOVIRUS
Physalis vein blotch	possibly luteovirus
PHYTOREOVIRUS GROUP[63]	A genus of the Reoviridae. Icosahedral particles of 75 nm diameter (510 S). The 59-nm-core carries 12 double spikes, through which the RNA is extruded during viral replication. It consists of 12 double-stranded RNA molecules of $0.3-3.0 \times 10^6$ daltons (total 16×10^6, 22% of particle weight). The virion is made up of seven of the 12 proteins coded for by the RNAs ($35-160 \times 10^3$ daltons). Wound tumor virus is the type member and rice dwarf and possibly rice gall dwarf viruses belong to this group. The host range is narrow; vectors are specific leafhoppers, in which the viruses also replicate.
PHYTORHABDOVIRUS GROUP[64]	This term for plant rhabdoviruses has not (yet) been approved by the ICTV. It is analogous to the approved term phytoreovirus. The virions are baccilliform or more rarely bullet-shaped, of $45-95 \times 135-380$ nm. The helical nucleocapsid consists of a minus-strand RNA of about 4×10^6 daltons and protein N (57×10^3 daltons). Subgroup A (type member lettuce necrotic yellow virus) resembles in its properties the vesiculoviruses, while subgroup B

Virus	Classification or properties (Group names in Capitals)
PHYTORHABDOVIRUS GROUP[64] (*continued*)	(type member potato yellow dwarf virus) shows similarities to the lyssaviruses. The gene order is 5' L-G-M-NS N 3', the proteins about 160, 85, 21, 45, 57 × 10³ daltons (RNA polymerase, glycoprotein, matrix, nonstructural, and nucleocapsid protein). The viruses are mechanically transmissible, but in nature, most of them are transmitted by aphids, some by leafhoppers; the viruses replicate in their vectors. Each subgroup has four members, but there are many probable additional phytorhabdoviruses.
Pigeon pea proliferation	possibly phytorhabdovirus (Rhabdoviridae)
Pigweed mosaic	possibly phytorhabdovirus (Rhabdoviridae)
Pineapple chlorotic beefstreak	possibly phytorhabdovirus (Rhabdoviridae)
Pisum	probably phytorhabdovirus (Rhabdoviridae)
Pittosporum vein yellowing	probably phytorhabdovirus (Rhabdoviridae)
Plantago	possibly caulimovirus (double-stranded DNA)
Plantago lanceolata: *see* plantain mottle virus	
Plantago mottle	TYMOVIRUS, possibly strain of scrophularia mottle
Plantago severe mottle	POTEXVIRUS
Plantago X: *see* plantain X virus	
Plantain mottle	possibly phytorhabdovirus (Rhabdoviridae)
Plantain X	POTEXVIRUS
Plum line pattern	synonym for apple mosaic virus
Plum pox	POTYVIRUS
PM 1–6	defective strains of tobacco mosaic virus
Poa semilatent	HORDEIVIRUS
Pod mottle	synonym for bean pod mottle virus

Virus	Classification or properties (Group names in Capitals)
Poinsettia cryptic	cryptic viruses
Poinsettia mosaic	possibly tymovirus
Pokeweed mosaic	POTYVIRUS
Poplar latent	CARLAVIRUS
Poplar mosaic	CARLAVIRUS
Potato A	POTYVIRUS
Potato acropetal necrosis	synonym for potato Y (requires helper virus for aphid transmission)
Potato aucuba mosaic	POTEXVIRUS
Potato black ringspot	NEPOVIRUS
Potato calico	identical with potato aucuba mosaic virus
Potato corkyring	strain of tobacco rattle virus
Potato E	identical with potato M virus
Potato F	identical with potato aucuba mosaic virus
Potato G	identical with potato aucuba mosaic virus
Potato interveinal mosaic	identical with potato M virus
Potato latent	POTEXVIRUS
Potato leaf roll	LUTEOVIRUS
Potato M	CARLAVIRUS
Potato mild mosaic	POTEXVIRUS
Potato mop top[65]	Possibly tobamovirus (stable rods of 100–150 and 250–300 nm (pitch 2.4 nm, 20 × 10^3 dalton protein). Serologically distantly related to tobacco mosaic virus, transmitted by fungus (*Spongosporum subterraneum*), narrow host range. Much similarity to beet necrotic yellow vein and soilborne beet necrotic yellow vein mosaic virus.
Potato P	identical with potato A virus
Potato paracrinkle	identical with potato M virus
Potato phloem necrosis	identical with potato leaf roll virus

Virus	Classification or properties (Group names in Capitals)
Potato S	CARLAVIRUS
Potato spindle tuber	type member of VIROIDS
Potato stem mottle	strain of tobacco rattle virus
Potato T	proposed member of proposed capillovirus group
Potato X	type member of POTEXVIRUS group (Figure 14 and 20)
Potato Y	type member of POTYVIRUS group
Potato yellow dwarf	PHYTORHABDOVIRUS (RHABDOVIRIDAE), type member of subgroup B (leafhopper-transmitted)
POTEXVIRUS GROUP[66]	Filamentous, comparatively stable particles of 13 × 470–580 nm (114–120 S, density in CsCl 1.31 g/cm^3) consisting of an RNA (6%) of 2.3 × 10^6 and 1000–1500 protein molecules of 14.3 × 10^3 daltons (helical pitch 3.4 nm). The protein becomes easily degraded so that lower molecular weights are suspect. The 5' end of the RNA is capped (see alfalfa mosaic virus); poly(A) is not at the 3' end, but the high content of A (32%) suggests internal poly(A) sequences. Potato virus X is the type member and at least 16 other members and many possible ones are known. Serological interrelationships are frequent, but not close. The host ranges are usually narrow. Most potexviruses produce characteristic banded cellular inclusion bodies (Fig. 20).
POTYVIRUS GROUP[67]	A very numerous group of filamentous viruses, 11 × 720–770, and in some instances up to 900 nm (about 150 S, density in CsCl 1.31 g/cm^3). The nucleoprotein helix has a pitch of 3.4 nm, the protein is about 33 × 10^3 daltons, the RNA 3.3 × 10^6 daltons and always about 5% of the particle weight. Most potyviruses are seed-borne, and transmitted by aphids in non-persistent manner. They form typical cytoplasmic inclusion bodies ("pinwheels"). Most of them show serological relation-

Virus	Classification or properties (Group names in Capitals)
POTYVIRUS GROUP[67] (*continued*)	ship to one or more other potyviruses, but not to other filamentous viruses. Their host range is usually narrow.
Primula mosaic	possibly potyvirus (aphid-transmitted)
Prune dwarf	ILARVIRUS
Prunus necrotic ringspot	ILARVIRUS
PSV-F, PSV-S	MYCOVIRUS (double-stranded RNA)
Quail pea mosaic	COMOVIRUS
Radish mosaic	COMOVIRUS
Radish P and R	synonym for turnip mosaic virus
Radish yellow edge	possibly cryptic virus (30 nm isometric, 113 S, seed-borne)
Radish yellows	synonym for beet western yellows
Ranunculus rapens symptomless	possibly phytorhabdovirus (Rhabdoviridae)
Raphanus	probably phytorhabdovirus (Rhabdoviridae)
Raspberry bushy dwarf	ILARVIRUS
Raspberry leaf curl	possibly luteovirus
Raspberry ringspot	NEPOVIRUS
Raspberry vein chlorosis	probably phytorhabdovirus (Rhabdoviridae)
Raspberry yellow dwarf	synonym for arabis mosaic
Raspberry yellow mosaic	synonym for rubus yellow net
Ratel	synonym for tobacco rattle virus
Rayado fino: *see* maize rayado fino	
Red clover mosaic	CARLAVIRUS
Red clover mosaic	possibly phytorhabdovirus (Rhabdoviridae)
Red clover mottle	COMOVIRUS
Red clover necrotic mosaic	DIANTHOVIRUS

Virus	Classification or properties (Group names in Capitals)
Red clover vein mosaic	CARLAVIRUS
Red current ringspot	synonym for raspberry ringspot virus
Reed canary mosaic	possibly potyvirus (aphid-borne)
REOVIRIDAE	The family encompassing the animal virus genera REOVIRUS, ORBIVIRUS, and ROTAVIRUS and the plant virus group PHYTORHABDO-VIRUS and FIJIVIRUS.
Rhabarber–Mosaik	synonym for arabis mosaic virus
RHABDOVIRIDAE	The family encompassing the animal virus genera VESICULOVIRUS and LYSSAVIRUS and the PLANT (PHYTO)RHABDOVIRUSES.
Rhododendron necrotic ringspot	possibly potexvirus
Rhubarb I, II	possibly potexvirus
Ribgrass mosaik: *see* Holmes ribgrass mosaic virus	
Rice black streaked dwarf	FIJIVIRUS (REOVIRIDAE), (closely related sero-logically to maize rough dwarf virus, leaf-hopper-borne)
Rice dwarf	PHYTOREOVIRUS (Reoviridae), (leafhopper-borne)
Rice gall dwarf	PHYTOREOVIRUS (REOVIRIDAE)
Rice giallume	LUTEOVIRUS
Rice grassy stunt	possibly maize rayado fino group
Rice hoja blanca	possibly rice stripe virus group
Rice mosaic	synonym for rice dwarf virus
Rice necrosis mosaic	possibly potyvirus (fungus-borne)
Rice necrosis mosaic	possibly potyvirus (fungus-borne)
Rice penyakit habang	related to cacao swollen shoot
Rice ragged stunt[68]	probably Fijivirus (Reoviridae), (but eight RNA segments)
RICE STRIPE VIRUS GROUP[69]	type member including maize stripe virus; 8 nm filaments of varying lengths at times branched, containing supercoiled circular nucleoprotein threads 3–4-nm wide and of

Virus	Classification or properties (Group names in Capitals)
RICE STRIPE VIRUS GROUP[69] (*continued*)	various lengths, containing 4 or 5 RNAs of $0.5-3 \times 10^6$ daltons and a protein of 32×10^3 daltons; transmitted by and replicating in leafhoppers).
Rice stunt	synonym for rice dwarf virus
Rice transitory yellowing	probably phytorhabdovirus (Rhabdoviridae), (leafhopper-borne)
Rice tungro bacilliform[70]	unclassified (31 × 100 nm, leafhopper-transmitted, not serologically related to other viruses)
Rice tungro spherical[43]	unclassified (30 nm diameter particles, 175 S, containing RNA of 3.2×10^6 daltons, leafhopper transmitted)
Rice yellow mottle	possibly sobemovirus (beetle-borne)
Rice yellow orange leaf	identical with spherical rice tungro virus
Robinia mosaic	unclassified (40 nm isometric particles, aphid-transmitted, wide host range)
Roll–Mosaik	identical to potato M virus
Rose infectious chlorosis	synonym of apple mosaic virus
Rose mosaic	ILARVIRUS (possibly identical to apple mosaic virus)
Rotkleeader–Mosaik: *see* red clover vein mosaic virus	
Rubus yellow net	probably phytorhabdovirus (Rhabdoviridae), (transmitted by aphids, not mechanically)
Russian winter wheat mosaic	probably phytorhabdovirus (Rhabdoviridae), (leafhopper-transmitted)
Ryegrass	probably phytorhabdovirus (Rhabdoviridae), (leafhopper-transmitted)
Ryegrass mosaic	possibly potyvirus (mite-transmitted, no serological relationship to potyviruses)
Ryegrass spherical	cryptic virus
Saccharomyces spp.	MYCOVIRUS (double-stranded RNA)

Virus	Classification or properties (Group names in Capitals)
Saguaro cactus[71]	possibly tombusvirus (structurally similar to tombusvirus, but no serological relationship to any virus detected)
Saintpaulia leaf necrosis	possibly phytorhabdovirus (Rhabdoviridae)
Sambucus vein clearing	possibly phytorhabdovirus (Rhabdoviridae)
Sammon's Opuntia	TOBAMOVIRUS (structurally similar to tobacco mosaic virus)
Sarracenia purpurea	possibly potyvirus
Satellite tobacco necrosis (STNV)[72]	Monotypic: 17 nm icosahedral particles (50 S), consisting of 0.4×10^6 dalton RNA (sequenced) and a protein of 23×10^3 daltons. Defective virus, the RNA coding only for its coat protein, unrelated yet specifically dependent on helper tobacco necrosis virus (even strain-specific dependency).
SATELLITE VIRAL RNAS[73]	Many viruses, particularly the nepoviruses and the cucumoviruses, carry at times one, several, or many nongenomic RNAs of $0.1–0.5 \times 10^6$ daltons in their virions. The function of these RNAs varies; in many instances no function is known, at times they affect the severity of symptoms.
Satsuma dwarf	possibly nepovirus (no known vector, seed-transmitted, no serological relationship detected)
Sawgrass	probably phytorhabdovirus (Rhabdoviridae)
Scrophularia mottle	TYMOVIRUS
ScV1, ScV2	MYCOVIRUS (double-stranded RNA)
Severe etch: *see* tobacco etch virus	
Shallot latent	CARLAVIRUS
Sharka plum	POTYVIRUS (serologically related to pepper veinal mottle and Datura shoestring virus)

Virus	Classification or properties (Group names in Capitals)
Sobemovirus group[74]	Icosahedral 27–30-nm-diameter particles (115 S) consisting of an RNA of 1.4×10^6 and 180 molecules of a coat protein of about 30×10^3 daltons. The RNA has a small 5' terminal protein that is essential for infectivity, and no characteristic 3' terminal features. Southern *bean mosaic* virus (so be mo) is the type member and turnip rosette virus and possibly others belong to this group. Each has a narrow host range. Transmission is by seeds and by beetles. No serological relationships detected.

Soil-borne wheat mosaic: *see* wheat soilborne mosaic virus

Solanum 1,2,3	synonyms for potato X, Y, A
Solanum 7,11	synonym for potato M
Solanum nodiflorum mottle[75]	possibly sobemovirus (1.5×10^6 dalton RNA and 0.12×10^6 dalton circular satellite RNA)
Solanum yellows	Luteovirus
Sonchus	Phytorhabdovirus (Rhabdoviridae), (subgroup A)
Sonchus yellow net	Phytorhabdovirus (Rhabdoviridae), (subgroup B), (alphid-transmitted)
Sorghum stunt mosaic	probably phytorabdovirus (Rhabdoviridae), (aphid-transmitted)
Sour cherry necrotic ringspot	similar to tobacco streak virus
Sour cherry yellow	synonym for prune dwarf virus
Southern bean mosaic[76]	type member of sobemo virus group
Sowbane mosaic	possibly sobemovirus (no serological relationship to other viruses)
Sowthistle yellow vein	Phytorhabdovirus (Rhabdoviridae), (subgroup B, aphid-transmitted)
Soybean dwarf	Luteovirus

Virus	Classification or properties (Group names in Capitals)
Soybean mosaic	POTYVIRUS
Spartina mottle	possibly potyvirus (mite-borne)
Spinach blight	synonym for cucumber mosaic virus
Spinach latent	ILARVIRUS
Spinach yellow dwarf	unclassified (possibly tobamovirus) (rigid rods of 15 × 250 nm, narrow host range)
Squach mosaic	COMOVIRUS
Star mottle	synonym for sowbane mosaic virus
Statice Y	possible potyvirus (aphid-transmitted)
Strawberry crinkle	PHYTORHABDOVIRUS (RHABDOVIRIDAE), (aphid-borne)
Strawberry latent ringspot[77]	possibly nepovirus (but two proteins of 44 and 29×10^3 and two RNAs to 2.4 and 1.4 daltons and, at times, a satellite RNA of 0.4×10^6 daltons that codes for a protein; no serological relationship to others)
Strawberry mild yellow edge	possibly luteovirus
Strawberry vein banding	CAULIMOVIRUS (double-stranded DNA)
Strongwellsea magna fungus	possibly Baculoviridae
Subterranean clover mottle	possibly sobemovirus with circular satellite RNA of 0.12×10^6 daltons
Subterranean clover red leaf	LUTEOVIRUS
Subterranean clover stunt	probably luteovirus
Sugar beet yellows	CLOSTEROVIRUS
Sugarcane Fiji disease	FIJIVIRUS (REOVIRIDAE), (leaf-hopper-transmitted)
Sugarcane mosaic	POTYVIRUS
Sunblotch	VIROID

Virus	Classification or properties (Group names in Capitals)
Sunflower ringspot	similar to alfalfa mosaic virus but not serologically related
Sunnhemp mosaic[78]	TOBAMOVIRUS (coat protein gene separately encapsidated in 40-nm long rod, distantly related to tobacco mosaic virus)
Sweet clover necrotic mosaic	DIANTHOVIRUS (soil-borne)
Sweet potato A	possibly potyvirus (aphid-transmitted)
Sweet potato mild mottle	possibly potyvirus (wildfly-transmitted)
Sweet potato russet crack	possibly potyvirus (aphid-transmitted)
Tabak Kräusel Krankheit	strain of tobacco rattle virus
Tabakmauche	strain of tobacco rattle virus
Tabak–Streifen	strain of tobacco rattle virus
Tamarillo mosaic	*Potyvirus*
Teasel mosaic	possibly potyvirus (aphid-transmitted)
Tephrosia symptomless	possibly tombusvirus (but no serological relationship to any other virus detected, narrow host range)
Theobroma 1 (or inflans)	synonym for cocoa swollen shoot
Thielaviopsis basicola	MYCOVIRUS (double-stranded RNA)
Thistle mottle	possibly caulimovirus
Thraustochytrium	MYCOVIRUS (double-stranded RNA)
Tobacco etch	POTYVIRUS (with 5′ terminal protein)
Tobacco leaf curl	probably geminivirus (possibly identical with tomato yellow dwarf)
Tobacco mosaic (TMV)[79]	TOBAMOVIRUS (type member of group). The TMV rods are 18×300 nm, helical pitch 2.3 nm (194 S), density in CsCl 1.325 g/cm^3. The RNA has a molecular weight of 2×10^6. It is 5′ terminally capped, and

Virus	Classification or properties (Group names in Capitals)
Tobacco mosaic (TMV)[79] (continued)	its 3′ end accepts histidine under tRNA charging conditions. Of the coat protein of 158–161 amino acids in different strains (17.5–18 × 10^3 daltons) 2130 molecules form the typical rod, each bound to three nucleotides of the RNA (4 nm radius), thus 6390 make up the total length of the viral RNA; it and the coat protein, as well as that of the proteins of several of its strains have been sequenced.
Tobacco necrosis[80]	type member of NECROVIRUS group which includes cucumber necrosis and melon necrotic spot virus: Icosahedral 28 nm diameter particles (118 S, density in CsCl 1.4 g/cm³) consisting of 1.4 × 10^6 dalton RNA and 180 protein molecules of 22.16 × 10^3 daltons; soil-borne, transmitted by fungus (Olpidius), wide host range among angiosperms, not serologically related to other viruses tested including cucumber necrosis virus. Often associated with satellite virus (satellite tobacco necrosis virus) (Figure 22)
Tobacco necrotic dwarf	LUTEOVIRUS
Tobacco rattle[81]	type member of TOBRAVIRUS group (Figure 23)
Tobacco ringspot[82]	type member of NEPOVIRUS group (often containing multiple molecules of satellite RNA of 0.1 × 10^6 daltons)
Tobacco rosette	unclassified (apparently a complex of two, possibly defective viruses)
Tobacco stem tumor	possibly geminivirus (icosahedral 17 nm diameter particles, seed-transmitted, wide host range)
Tobacco streak[83]	type member of ILARVIRUS group
Tobacco vein assistor	possibly luteovirus
Tobacco vein distorting	possibly luteovirus
Tobacco vein mottling	POTYVIRUS (aphid-borne)
Tobacco yellow dwarf	probably geminivirus (single-stranded DNA)

Virus	Classification or properties (Group names in Capitals)
Tobacco yellow net	possibly luteovirus
TOBAMOVIRUS GROUP [84]	A most intensively studied relatively small group of very stable rigid rod virions. *Tobacco mosaic* is its type member (U1).

TOBAMOVIRUS GROUP [84]

A most intensively studied relatively small group of very stable rigid rod virions. *Tobacco mosaic* is its type member (U1).

Other members are cucumber green mottle mosaic (CGMMV), cucumber 4 (CV4), Frangipani mosaic, Odontoglossum ringspot, (Holmes) ribgrass mosaic (HR), (sequenced protein), Sammon's Opuntia, sunnhemp mosaic (sequenced protein), tomato mosaic, and U2-tobacco mosaic viruses. Dahlemense, masked, and green and yellow aucuba viruses are strains of U1. All of these are serologically more or less closely related. Their host ranges vary from wide to limited. Transmission is generally only mechanical.

Other rigid rod viruses of varying or multiple lengths and stabilities, regarded as possible members on the basis of their shape, are broad bean necrosis, beet necrotic yellow vein, chara corallina, Nicotiana velutina mosaic, peanut clump, potato mop-top, and soil-borne wheat mosaic virus. For chara corallina and potato mop-top distant serological relationships to U1 have been reported.

TOBRAVIRUS GROUP [85]

Composed of two types of tubular helical particles (pitch 2.5 nm) of 22 nm diameter. The longer particle (L) is about 200-nm long and the shorter ranges from 46–114 nm length for different strains of viruses (300 S and 150–250 S, density in CsCl 1.30–1.32 g/cm^3). The coat protein is about 22×10^3 daltons. The RNAs (5%) of the two types of particles are 2.4 and $0.6–1.4 \times 10^6$ daltons. Only the latter RNA is capped. In contrast to other multiparticle viruses, that require several components for infectivity, the larger RNA or virion of tobraviruses is by itself infectious. However, no virions are pro-

Virus	Classification or properties (Group names in Capitals)
TOBRAVIRUS GROUP[85] (continued)	duced as the result of such infection, since the coat protein gene is located on the smaller RNA. The type member is tobacco rattle virus, and the only known other members are pea early browning virus, and possibly peanut clump virus. Serological relations vary greatly even within strains of the type members. The host range is wide, transmission by nematodes in persistent manner and by seed.
Tollkirschen–Scheckungs	synonym for belladonna mottle virus
Tomato aspermy[86]	CUCUMOVIRUS
Tomato atypical mosaic (Y,G)	strains of tobacco mosaic virus (Y = yellow, G = green)
Tomato black ring[87]	NEPOVIRUS (with satellite RNA 3 of 0.5×10^6 daltons)
Tomato bunchy top	synonym for potato spindle tuber viroid
Tomato bushy stunt	type member of TOMBUSVIRUS group
Tomato etch	probably identical with tobacco etch virus
Tomato fernleaf	synonym for cucumber mosaic virus
Tomato leaf curl	GEMINIVIRUS (possibly identical with tomato yellow mosaic and leaf curl), (single-stranded DNA)
Tomato (golden) mosaic	GEMINIVIRUS (possibly synonym for tomato yellow dwarf), (single-stranded DNA), serologically related to cassava latent virus
Tomato mosaic	TOBAMOVIRUS (serologically related to tobacco mosaic virus, 30 amino acid exchanges, host range wide among Solanaceous species)
Tomato Peru mosaic	probably potyvirus (transmitted by aphids)
Tomato ringspot	NEPOVIRUS
Tomato spotted wilt[88]	Monotypic: 82 nm spherical particles (560 S, density 1.21 g/cm^3 in sucrose, probably

Virus	Classification or properties (Group names in Capitals)
Tomato spotted wilt[88] (*continued*)	plus-strand RNAs lacking poly(A) of 2.6, 1.9, 1.7, and 1.3×10^6 daltons, four glycoproteins ranging from 78 to 27×10^3 daltons, 20% lipid, wide host range, transmitted by thrips in persistent manner)
Tomato top necrosis	possibly nepovirus
Tomato yellow dwarf	GEMINIVIRUS, possibly the same as tomato leaf curl and tomato (golden) mosaic virus (single-stranded DNA)
Tomato yellow net and top	possibly and probably luteoviruses
TOMBUSVIRUS GROUP[89]	Icosahedral particles of 30 nm diameter (140 S). These consist of one molecule of RNA (17%) of 1.5×10^6 and a coat protein of 41×10^3 daltons.\n\n*Tomato bus*hy stunt virus is the type member, and others are artichoke mottled crinkle, carnation Italian ringspot, Cymbidium ringspot, eggplant mottled crinkle, Pelargoinium leaf curl, and Petunia asteroid mosaic virus. All are serologically closely related.\n\nThe host range is wide, transmission mechanical and through the soil.
Tradescantia/Zebrina	possibly potyvirus (aphid-borne)
Transitory yellowing: *see* rice transitory yellowing	
Trespen–Mosaik	synonym for brome mosaic virus
Tricorna viruses[90]	A proposed name for the group of plant viruses with tripartite single-stranded RNA genome
Trifolium 2	synonym for red clover vein mottle virus
Triticum aestivum chlorotic spot	possibly phytorhabdovirus (Rhabdoviridae)
True broad bean mosaic: *see* broad bean true mosaic virus	
Tulare apple mosaic	ILARVIRUS
Tulare apple streak mosaic	ILARVIRUS

Virus	Classification or properties (Group names in Capitals)
Tulip Augusta disease	identical with tobacco necrosis virus
Tulip breaking	POTYVIRUS (narrow host range)
Tulip mosaic	synonym for tulip-breaking virus
Tungro: *see* rice tangro virus	
Turnip 2	strain of radish mosaic virus
Turnip crinkle[91]	structurally similar to tombusvirus (no serological relationship to other viruses, transmitted by flee-beetle). Virion contains a satellite RNA of 0.17×10^6 daltons.
Turnip mosaic	POTYVIRUS
Turnip rosette	SOBEMOVIRUS
Turnip yellow mosaic[92]	type member of TYMOVIRUS group (Figure 24)
Turnip yellow	LUTEOVIRUS
TYMOVIRUS GROUP[93]	Icosahedral particles of 29 nm diameter (115 S, density in CsCl 1.42 g/cm³), composed to 35% of RNA of 2×10^6, a coat protein mRNA of 0.25×10^6 daltons, of known sequence and identical to that of the 3' end of the genomic RNA (frequently also larger subgenomic RNA molecules), and 180 molecules of the coat protein of 20×10^3 daltons. The RNAs are high in cytidine (up to 41%); they are 5' terminally capped and the 3' terminus accepts valine under tRNA charging conditions. "Top component" particles lacking RNA (54 S) always accompany the virions. The type member is *turnip yellow mosaic* virus (tymo), and there are at least 16 other members of the group. All are more or less closely serologically related, and several groups so closely that they may be regarded as strains, rather than separate viruses, e.g., Andean potato latent, belladonna mottle, dulcamara mottle, eggplant mosaic, and physalis mosaic viruses.

Virus	Classification or properties (Group names in Capitals)
TYMOVIRUS GROUP[93] (*continued*)	Their host ranges are usually narrow (almost only dicotyledous), and transmission is mechanical and by beetles.
U1	Tobamovirus, common tobacco mosaic virus (vulgare)
U2–U8	strains of tobacco mosaic virus (U6 closely related to U1, U2 less closely related)
UmV (*Ustilago maidis*)	unclassified
Urdbean leaf crinkle	unclassified, similar to pea enation mosaic virus
Ustilago maydis	MYCOVIRUSES (double-stranded RNA)
UW21, UW51 (*Ustilago maidis*)	unclassified
Velvet tobacco mottle[94]	possibly sobemovirus with 0.12×10^6 dalton circular satellite RNA
Vicia cryptic	cryptic viruses
Vigna sinensis mosaic	possibly phytorhabdovirus (Rhabdoviridae)
Viola mottle	POTEXVIRUS
VIROIDS[95]	An increasing number of plant diseases is found to be caused by small cyclic single-stranded RNA molecules. Most of these are about 360 nucleotides long, and most have been sequenced. They are characterized by a specific rod-shaped conformation caused by hairpin looping, and unaffected by considerable differences in nucleotide sequences. Neither viroids nor their complementary RNAs are translated. Their mode of transcription–replication and infectivity remains unknown. The type member is potato spindle tuber viroid, and others are citrus exocortis, chrysanthemum stunt, chrysanthemum chlorotic mottle viroid, etc.
Voandzeia mosaic	CARLAVIRUS, possibly strain of cowpea mild mottle virus, though different host range

Virus	Classification or properties (Group names in Capitals)
Watermelon mosaic 1 and 2	POTYVIRUSES
Weidelgrass–Mosaik	synonym for brome mosaic
Wheat 1 and 3	synonym for wheat mosaic
Wheat 6 and 7	synonym for wheat streak mosaic
Wheat chlorotic streak	probably phytorhabdovirus (Rhabdoviridae), (leaf-hopper transmitted)
Wheat dwarf	probably geminivirus
Wheat (soil-borne) mosaic[96]	Possibly tobamovirus (rods of 281, 138, and 92 nm, RNA of 2.3, 1.2, and 1.0×10^6 daltons, protein of 19.7×10^3 daltons). Either of the shorter rods plus the long one needed for infectivity. Distant serological relationship to tobacco mosaic virus, transmitted by fungus (*Polymyxa graminis*), much similarity to beet necrotic yellow vein and potato mop-top viruses.
Wheat spindle streak	POTYVIRUS (aphid-transmitted)
Wheat spindle streak mosaic	possibly closterovirus (up to 2000 nm long, fungus-transmitted)
Wheat streak	POTYVIRUS (aphid-transmitted)
Wheat streak mosaic	POTYVIRUS (mite-transmitted, no serological relationships found)
Wheat striate mosaic	PHYTORHABDOVIRUS (RHABDOVIRIDAE), subgroup A (leafhopper-transmitted)
Wheat take-all	MYCOVIRUS (double-stranded RNA)
Wheat yellow leaf	CLOSTEROVIRUS, subgroup I
Wheat yellow mosaic	possibly potyvirus (fungus-transmitted)
White bryony mosaic	possibly carlavirus (aphid-transmitted)
White clover	synonym for cymbidium ringspot virus
White clover mosaic	POTEXVIRUS
Wild cucumber mosaic	TYMOVIRUS
Wild potato mosaic	possibly potyvirus (aphid-transmitted)

Virus	Classification or properties (Group names in Capitals)
Wineberry latent	possibly potexvirus (but no serological relationship to others tested)
Winter wheat mosaic: *see* Russian winter wheat mosaic virus	
Wisconsin pea stunt	synonym for red clover vein mosaic virus
Wisteria vein mosaic	POTYVIRUS
Wound tumor[97]	type member of PHYTOREOVIRUS group (REOVIRIDAE) (Figure 25)
Yam internal brownspot	related to cacao swollen shoot
Yam mosaic	identical with Dioscorea green-banding virus
Yellow aucuba: *see* aucuba virus	
Yellow bean mosaic: *see* bean mosaic virus	
Yellow cucumber mosaic	CUCUMOVIRUS
Yellow mottle	unclassified 25-nm-diameter beetle-borne rice virus
Yellow spot mosaic	strain of alfalfa mosaic virus
Yucai	TOBAMOVIRUS
Zea (mays): *see* maize mosaic virus	
Zygocactus (X)	possibly potexvirus
01-1-4-A,0-1916-A,3 bla-A, 3 bla-B,3 bla-C,38-4-A 4919-A	MYCOVIRUS (double-stranded RNA)

References (Section II)

1a. Vloten-Doting, L. van, and Jaspars, E. M. J., 1979, Plant covirus systems: Three-component systems, *Comp. Virol.* **11:**1.

1b. Nassuth, A., Alblas, F., and Bol, J. F., 1981, Localization of genetic information involved in the replication of alfalfa mosaic virus, *J. Gen. Virol.* **53:**207–214.

1c. Koper-Zwarthoff, E. C., and Bol, J. F., 1980, Nucleotide sequence of the putative recognition site for coat protein in the RNAs of alfalfa mosaic virus and tobacco streak virus, *Nucl. Acids. Res.* **8:**3307.

1d. Gould, A. R., and Symons, R. H., 1978, Alfalfa mosaic virus RNA. Determination of the sequence homology between the four RNA species and a comparison with the four RNA species of cucumber mosaic virus, *Eur. J. Biochem.* **91:**269–278.

1e. Tol, R. G. van, and Vloten-Doting, L. van, 1981, Lack of serological relationship between the 35K nonstructural protein of alfalfa mosaic virus and the corresponding proteins of three other plant viruses with a tripartite genome, *Virology* **109:**444–447.

2a. Dodds, J. A., and Cole, A., 1980, Microscopy and biology of *Uronema gigas*, a filamentous eukaryotic green alga, and its associated tailed virus-like particle, *Virology* **100:**156–165.

2b. Cole, A., Dodds, J. A., and Hamilton, R. I., 1980, Purification and some properties of a double-stranded DNA containing virus-like particle from *Uronema gigas*, a filamentous eukaryotic green alga, *Virology* **100:**166–174.

2c. Stanker, L. H., Hoffman, L. R., and MacLeod, R., 1981, Isolation and partial chemical characterization of a virus-like particle from a eukaryotic alga, *Virology* **114:**357–369.

3. Jones, R. A. C., and Kenton, R. H., 1980, CMI/AAB, *Descriptions of Plant Viruses*, Commonwealth Mycological Institute and Association of Applied Biologists, No. 216.

4. Kenton, R. H., and Jones, R. A. C., 1983, CMI/AAB, *Descriptions of Plant Viruses*, Commonwealth Mycological Institute and Association of Applied Biologists, No. 270.

5a. Boykov, S. V., Taliansky, M. E., Malyshenko, S. I., Kozlov, Yu V., and Atabekov, J. G., 1981, Homology in the 3'-terminal regions of different barley stripe mosaic virus RNA species, *Virology* **113:**168–173.

5b. Agranovsky, A. A., Dolja, V. V., and Atabekov, J. G., 1983, Differences in polyadenylate length between individual barley stripe mosaic virus RNA species, *Virology* **129:**344–349.

5c. Jackson, A. O., and Brakke, M. K., 1973, Multicomponent properties of barley stripe mosaic RNA, *Virology* **55:**483.

6. Rochow, W. F., Aapola, A. I. E., Brakke, M. K., and Carmichael, E., 1971, Purification and antigenicity of barley yellow dwarf virus, *Virology* **46:**117.

7. Waterworth, H., 1981, CMI/AAB, *Descriptions of Plant Viruses*, Commonwealth Mycological Institute and Association of Applied Biologists, No. 231.

8a. Steven, A. C., Trus, B. L., Putz, C., and Wurtz, M., 1981, The molecular organization of beet necrotic yellow vein virus, *Virology* **113:**428–438.

8b. Putz, C., 1977, Composition and structure of beet necrotic yellow vein virus, *J. Gen. Virol.* **35:**397–401.

9. Scott, H. A., and Hoy, J. W., 1981, CMI/AAB, *Descriptions of Plant Viruses*, Commonwealth Mycological Institute and Association of Applied Biologists, No. 237.

10. Inouye, T., and Nakasone, W., 1980, CMI/AAB, *Descriptions of Plant Viruses*, Commonwealth Mycological Institute and Association of Applied Biologists, No. 223.

11. Doel, T. R., 1975, Comparative properties of type, nasturtium ringspot and petunia ringspot strains of broad bean wilt virus, *J. Gen. Virol.* **26**:95–108.

12a. Matthews, R. E. F., 1982, Classification and nomenclature of viruses, Fourth Report of the International Commission on Taxonomy of Viruses, *Intervirology* **17**:173.

12b. Lane, L. C., 1979, Bromovirus group, CMI/AAB, No. 215.

12c. Dasgupta, R., Ahlquist, P., and Kaesberg, P., 1980, Sequence of the 3′ untranslated region of brome mosaic virus coat protein messenger RNA, *Virology* **104**:339–346.

13a. Matthews, R. E. F., 1982, Carlavirus group, *Intervirology* **17**:149.

13b. Koenig, A., 1982, Carlavirus group, CMI/AAB, No. 259.

13c. Veerisetty, V., and Brakke, M., 1977, Differentiation of legume carlaviruses based on their biochemical properties, *Virology* **83**:226.

14. Tremaine, J. H., 1970, Physical, chemical and serological properties of carnation mottle virus, *Virology* **42**:611.

15. Halk, E. L., Robinson, D. J., and Murant, A. F., 1979, Molecular weight of the infective RNA from leaves infected with carrot mottle virus, *J. Gen. Virol.* **45**:383–388.

16a. Matthews, R. E. F., 1982, Classification and nomenclature of viruses, Fourth Report of the International Commission on Taxonomy of Viruses, *Intervirology* **17**:64.

16b. Menissier, J., Lebeurier, G., and Hirth, L., 1982, Free cauliflower mosaic virus super-coiled DNA in infected plants, *Virology* **117**:322–328.

16c. Ansa, O. A., Bowyer, J. W., and Shepherd, R. J., 1982, Evidence for replication of cauliflower mosaic virus DNA in plant nuclei, *Virology* **121**:147–156.

16d. Hull, R., 1980, Structure of the cauliflower mosaic virus genome. III. Restriction endonuclease mapping of 33 isolates, *Virology* **100**:76–90.

16e. Lebeurier, G., Hirth, L., Hohn, B., and Hohn, T., 1982, *In vivo* recombination of cauliflower mosaic virus DNA, *Proc. Natl. Acad. Sci. USA* **79**:2932–2936.

17. Skotnicki, A., Gibbs, A., and Wrigley, N. G., 1976, Further studies on *Chara corallina* virus, *Virology* **75**:457.

18. Haseloff, J., and Symons, R. H., 1981, Chrysanthemum stunt viroid: Primary sequence and secondary structure, *Nucl. Acids Res.* **9**:2741.

19. Visvader, J. E., and Symons, R. H., 1983, Comparative sequence and structure of different isolates of citrus exorcortis viroid, *Virology* **130**:232–237.

20a. Matthews, R. E. F., 1982, Classification and nomenclature of viruses, *Intervirology* **17**:147.

20b. Bar-Joseph, M., and Murant, A. F., 1982, Closterovirus group, CMI/AAB, No. 260.

20c. Bar-Joseph, M., Garnsey, S. M., and Gonsalves, D., 1979, The closteroviruses: A distinct group of elongated plant viruses, *Adv. Virus Res.* **25**:93.

21. Huth, W., and Paul, H.-L., 1972, CMI/AAB, No. 107.

22. Matthews, R. E. F., 1982, Classification and nomenclature of viruses, *Intervirology* **17**:161.

23. Dawson, W. O., 1981, Effect of temperature-sensitive, replication-defective mutations on the synthesis of cowpea chlorotic mottle virus, *Virology* **115**:130–136.

24a. Franssen, H., Goldback, R., Broekhuijsen, M., Moerman, M., and Kammen, A. van, 1982, Expression of middle-component RNA of cowpea mosaic virus: *In vitro* generation of a precursor to both capsid proteins by a bottom-component RNA-encoded protease from infected cells, *J. Virol.* **41**:8–17.

24b. Najarian, R. C., and Bruening, G., 1980, Similar sequences from the 5′ end of cowpea mosaic virus RNAs, *Virology* **106**:301–309.

25. Bozarth, R. F., and Shoyinka, S. A., 1979, CMI/AAB, No. 212.

26. Beier, H., Issinger, O. G., Deuschle, M., and Mundry, K. W., 1981, Translation of the RNA of cowpea severe mosaic virus *in vitro* and in cowpea protoplasts, *J. Gen. Virol.* **54**:379–390.

27. Lisa, V., Boccardo, G., and Milne, R. G., 1981, Double-stranded ribonucleic acid from carnation cryptic virus, *Virology* **115**:410–413.

28a. Habili, N., and Francki, R. I. B., Comparative studies on tomato aspermy and cucumber mosaic viruses. I. Physical and chemical properties, *Virology* **57**:392–401.

28b. Yamaguchi, K., Hidaka, S., and Muira, K-I., 1982, Relationship between structure of the 5' noncoding region of viral mRNA and efficiency in the initiation step of protein synthesis in a eukaryotic system, *Proc. Natl. Acad. Sci. USA* **79**:1012–1016.

29. Dias, H. F., and McKeen, C. D., 1972, CMI/AAB, No. 82.

30. Tanne E., Sela, I., Assuline, I., Cohen, S., and Marco S., 1981, The etiology of isolation and characteristics of cucumber vein yellowing virus, Fifth International Congress of Virology Abstracts, p. 234.

31. Matthews, R. E. F., 1982, Classification and nomenclature of viruses, *Intervirology* **17**:171.

32. Mohamed, N. A., 1978, Physical and chemical properties of cynosurus mottle virus, *J. Gen. Virol.* **40**:379–389.

33. Matthews, R. E. F., 1982, Classification and nomenclature of viruses, *Intervirology* **17**:160.

34a. Matthews, R. E. F., 1982, Classification and nomenclature of viruses, *Intervirology* **17**:85.

34b. Hatta, T., and Francki, R. I. B., 1980, Morphology of Fiji disease virus, *Virology* **76**:797.

35a. Matthews, R. E. F., 1982, Classification and nomenclature of viruses, *Intervirology* **17**:76.

35b. Ikegami, M., Haber, S., and Goodman, R. M., 1981, Isolation and characterization of virus-specific double-stranded DNA from tissues infected by bean golden mosaic virus, *Proc. Natl. Acad. Sci. USA* **78**:4102–4106.

35c. Reisman, D., Ricciardi, R. P., and Goodman, R. M., 1979, The size and topology of single-stranded DNA from bean golden mosaic virus, *Virology* **97**:388.

36. Boccardo, G., and d'Aquilo, M., 1981, The protein and nucleic acid of a closterovirus isolated from a grapevine with stem-pitting symptoms, *J. Gen. Virol.* **53**:179–182.

37. Waterworth, H., 1980, CMI/AAB, No. 227.

38. Matthews, R. E. F., 1982, Classification and nomenclature of viruses, *Intervirology* **17**:178.

39a. Fulton, R. W., 1983, CMI/AAB, No. 275.

39b. Matthews, R. E. F., 1982, Classification and nomenclature of viruses, *Intervirology* **17**:175.

40. Falk, B. W., Morris, T. J., and Duffus, J. E., 1979, Unstable infectivity and sedimentable ds-RNA associated with lettuce speckless mottle virus, *Virology* **96**:239–248.

41. Matthews, R. E. F., 1982, Classification and nomenclature of viruses, *Intervirology* **17**:140.

42. Plěse, N., Koenig, R., Lesemann, R., and Bozarth, 1979, An elongated plant virus of uncertain classification, *Phytopathology* **69**:471.

43. Gingery, R. E., 1976, Properies of maize chlorotic dwarf virus and its ribonucleic acid, *Virology* **73**:311–318.

44. Gamez, R., 1980, CMI/AAB, No. 220.

45. Milne, G., and Lovisolo, O., 1977, Maize rough dwarf and related viruses, *Adv. Virus Res.* **21**:267.

46. Gingery, R. E., Nault, L. R., and Bradfute, O. E., Maize stripe virus: Characteristics of a member of a new virus class, *Virology* **112**:99–108.

47. Barton, R. J., and Hollings, M., 1981, Properties of melandrium yellowing of vein, Fifth International Congress of Virology Abstract, p. 235.

48. Barton, R. J., and Hollings, M., 1979, Purification and some properties of two viruses infecting the cultivated mushroom *Agaricus bisporus*, *J. Gen. Virol.* **42**:231–240.

49a. Hollings, M., 1978, Mycoviruses: Viruses that infect fungi, *Adv. Virus Res.* **22**:1.

49b. Lemke, P. A., and Nash, C. H., 1974, Fungal viruses, *Bacteriol. Rev.* **38**:29.

49c. Buck, K. W., Romanos, M. A., McFadden, J. J. P., and Rawlinson, C. J., 1981, *In vitro* transcription of double-stranded RNA by virion-associated RNA polymerases of viruses from *Gaeumannomyces graminis*, *J. Gen. Virol.* **57:**157–168.

50. Gallitelli, D., Piazzolla, P., Savino, V., Quacquarelli, A., and Martelli, G. P., 1981, A comparison of myrobalan latent ringspot virus with other nepoviruses, *J. Gen. Virol.* **53:**57–65.

51. Mowat, W. P., Asjet, C., and Brunt, A. A., 1976, CMI/AAB, No. 166.

52a. Harrison, B. D., and Murant, A. F., 1977, CMI/AAB, No. 185.

52b. Matthews, R. E. F., 1982, Classification and nomenclature of viruses, *Intervirology* **17:**163.

52c. Bruening, G., 1979, Plant covirus systems: Two-component systems, in: *Comprehensive Virology*, Vol. 2 (H. Fraenkel-Conrat and R. R. Wagner, eds.), p. 111, Plenum Press, New York.

52d. Mayo, M. A., Barker, H., and Harrison, B. D., 1979, Polyadenylate in the RNA of five nepoviruses, *J. Gen. Virol.* **43:**603–610.

52e. Harrison, B. D., Murant, A. F., and Mayo, M. A., 1972, Evidence for two functional RNA species in raspberry ringspot virus, *J. Gen. Virol.* **16:**339–348.

53. Randles, J. W., 1978, CMI/AAB, No. 189.

54. Banttari, E. E., and Zeyen, R. J., 1969, Chromatographic purification of oat blue dwarf virus, *Phytopathology* **59:**183.

55. Niblett, C. L., and Paulsen, A. Q., 1975, Purification and characterisation of panicam mosaic virus, *Phytopathology* **65:**1157.

56. Erickson, J. W., and Bancroft, J. B., 1978, The self-assembly of papaya mosaic virus, *Virology* **90:**36–46.

57. Murant, A. F., 1974, CMI/AAB, No. 130.

58. Reisman, D., and de Zoeten, G. A., 1982, A covalently-linked protein at the 5'-ends of the genomic RNAs of pea enation mosaic virus, *J. Gen. Virol.* **62;**187–190.

59. Thouvenel, J. C., and Fauguet, C., 1981, Further properties of peanut clump virus and studies on its natural transmission, *Ann. Appl. Biol.* **97:**99.

60. Hollings, M., 1974, CMI/AAB, No. 130.

61. Gallitelli, D., 1982, Properties of a tomato isolate of pelargonium zonate spot virus, *Ann. Appl. Biol.* **100:**457.

62. Paul, H.-L., Huth, W., and Querfurth, G., 1973–1974, Cocksfoot mild mosaic virus–phleum mottle virus: A comparison, *Intervirology* **2:**253.

63a. Matthews, R. E. F., 1982, Classification and nomenclature of viruses, *Intervirology* **17:**85.

63b. Francki, R. I. B., and Boccardo, G., 1983, The plant Reoviridae, in: *The Viruses*, The Reoviridae (W. Joklik, ed.), p. 505, Plenum Press, New York.

64a. Peters, D., 1981, CMI/AAB, No. 244.

64b. Dale, J. L., and Peters, D., 1981, Protein composition of the virions of five plant rhabdoviruses, *Intervirology* **16:**86.

65. Harrison, B. D., 1974, CMI/AAB, No. 138.

66a. Koenig, R., 1978, CMI/AAB, No. 200.

66b. Matthews, R. E. F., 1982, Classification and nomenclature of viruses, *Intervirology* **17:**156.

66c. Wodnar-Filipowicz, Al., Skrzeczkowski, L. J., and Filipowicz, W., 1980, Translation of potato virus X RNA into high molecular weight proteins, *FEBS Lett.* **109:**151.

66d. Richardson, J. F., Tollin, P., and Bancroft, J. B., 1981, The architecture of the potexviruses, *Virology* **112:**34–39.

67a. Hollings, M., and Brunt, A. A., 1981, CMI/AAB, No. 245.

67b. Matthews, R. E. F., 1982, Classification and nomenclature of viruses, *Intervirology* **17:**152.

67c. Abu-Samah, N., and Randles, J. W., 1981, A comparison of the nucleotide sequence homologies of three isolates of bean yellow mosaic virus and their relationship to other potyviruses, *Virology* **110:**436–444.

67d. Moghal, S. M., and Francki, R. I. B., 1981, Towards a system for the identification and classification of potyviruses, *Virology* **112:**210–216.

68. Milne, R. G., 1980, Does rice ragged stunt virus lack the typical double shell of the Reoviridae? *Intervirology* **14**:331.

69. Toriyama, S., 1982, Characterization of rice stripe virus: A heavy component carrying infectivity, *J. Gen. Virol.* **61**;187–195.

70. Galvez, G. E., 1968, Purification and characteristics of rice tungro virus by density–gradient configuration, *Virology* **35**:418.

71. Nelson, M. R., 1975, Physiochemical and serological properties of a virus from saguaro cactus, *Virology* **65**:309–319.

72a. Uyemoto, J. K., Grogan, R. G., and Wakeman, J. R., 1968, Selective activation of satellite virus strains by strains of tobacco necrosis virus, *Virology* **34**:410–418.

72b. Kaempfer, R., van Emmelo, J., and Fiers, W., 1981, Specific binding of eukaryotic initiation factor 2 to satellite tobacco necrosis virus RNA at a 5'-terminal sequence comprising the ribosome-binding site, *Proc. Natl. Acad. Sci. USA* **78**:1542–1546.

73a. Waterworth, H. E., Kaper, J. M., and Tousignant, M. E., 1979, CARNA 5, the small cucumber mosaic virus-dependent replicating RNA, regulates disease expression, *Science* **204**:845.

73b. Mossop, D. W., and Francki, R. I. B., 1979, The stability of satellite viral RNAs *in vivo* and *in vitro*, *Virology* **94**:243.

74. Matthews, R. E. F., 1982, Classification and nomenclature of viruses, *Intervirology* **17**:144.

75. Gould, A. R., and Hatta, T., 1981, Studies on encapsidated viroid-like RNA. III. Comparative studies on RNAs isolated from velvet tobacco mottle virus and *Solanum nodiflorum* mottle virus, *Virology* **109**:137.

76a. Rutgers, T., Salerno-Rise, T., and Kaesberg, P., 1980, Messenger RNA for the coat protein of Southern bean mosaic virus, *Virology* **104**:506–509.

76b. Argos, P., 1981, Secondary structure prediction of plant virus coat proteins, *Virology* **110**:55–62.

77. Murant, A. F., 1974, CMI/AAB, No. 126.

78. Higgins, T. J. V., Goodwin, P. B., and Whitefeld, P. R., 1976, Occurrence of short particles in beans infected with the cowpea strain of TMV. II. Evidence that short particles contain the cistron for coat-protein, *Virology* **71**:486.

79a. Goelet, P., Lomonossoff, G. P., Butler, P. J. G., Akam, M. E., Gait, M. J., and Karn, J., 1982, Nucleotide sequence of tobacco mosaic virus RNA, *Proc. Natl. Acad. Sci. USA* **79**:5818–5822.

79b. Bloomer, A. C. *et al.*, 1978, Protein disk of tobacco mosaic virus at 2.8 Å resolution showing the interactions within and between subunits, *Nature* **276**:362.

79c. Fraenkel-Conrat, H., and Singer, B., 1957, Virus reconstitution. II. Combination of protein and nucleic acid from different strains, *Biophys. Acta* **24**:540.

79d. Lebeurier, G., Nicolaieff, A., and Richards, K. E., 1977, Inside-out model for self-assembly of tobacco mosaic virus, *Proc. Natl. Acad. Sci. USA* **74**:149.

79e. Holmes, K. C., 1980, Protein–RNA interactions during assembly of tobacco mosaic virus, *Trends Biochem. Sci.* **5**:4.

79f. Pelham, H. R. B., 1978, Leaky UAG termination codon in tobacco mosaic virus RNA, *Nature* **272**:469.

79g. Hirth, L., and Richards, K. E., 1981, Tobacco mosaic virus: Model for structure and function of a simple virus, *Adv. Virus Res.* **26**:145.

80a. Matthews, R. E. F., 1982, Classification and nomenclature of viruses, *Intervirology* **17**:146.

80b. Kassanis, B., and Phillips, M. P., 1970, Serological relationship of strains of tobacco necrosis virus and their ability to activate strains of satellite virus, *J. Gen. Virol.* **9**:119–126.

81a. Mayo, M. A., 1982, Polypeptides induced by tobacco rattle virus during multiplication in tobacco protoplasts, *Intervirology* **17**:240.

81b. Sänger, H. L., 1968, Characteristics of tobacco rattle virus. I. Evidence that its two particles are functionally defective and mutually complementing, *Mol. Gen. Genet.* **101**:346.

82a. Kiefer, M. C., Daubert, S. D., Schneider, I. R., and Bruening, G., 1982, Multimeric forms of satellite of tobacco ringspot virus RNA, *Virology* **121**:262–273.

82b. Sogo, J. M., and Schneider, I. R., 1982, Electron microscopy of double-stranded nucleic acids found in tissue infected with the satellite of tobacco ringspot virus, *Virology* **117**:401–415.

82c. Schneider, I. R., and White R. M., 1976, Tobacco ringspot virus codes for the coat protein of its satellite, *Virology* **70**:244–246.

83a. Koper-Zwarthoff, E. C., and Bol, J. F., 1980, Nucleotide sequence of the putative recognition site for coat protein in the RNAs of alfalfa mosaic virus and tobacco streak virus, *Nucl. Acids Res.* **8**:3307.

83b. Vloten-Doting, L. van, 1975, Coat protein is required for infectivity of tobacco streak virus: Biological equivalence of the coat proteins of tobacco streak and alfalfa mosaic viruses, *Virology* **65**:215–225.

84a. Gibbs, A., 1980, How ancient are the tobamoviruses? *Intervirology* **14**:101.

84b. Gibbs, A. J., 1977, CMI/AAB, No. 184.

84c. Matthews, R. E. F., 1982, Classification and nomenclature of viruses, *Intervirology* **17**:158.

84d. Palukaitis, P., and Symons, R. H., 1980, Nucleotide sequence of 13 tobamovirus RNAs as determined by hybridization analysis with complementary DNA, *Virology* **107**:354–361.

85a. Harrison, B. D., and Robinson, D. J., 1978, The tobraviruses, *Adv. Virus Res.* **23**:25.

85b. Matthews, R. E. F., 1982, Classification and nomenclature of viruses, *Intervirology* **17**:170.

86. Wilson, P. A., and Symons, R. H., 1981, The RNAs of cucumoviruses: 3′-Terminal sequence analysis of two strains of tomato aspermy virus, *Virology* **112**:342–345.

87. Robinson, D. J., Barker, H., Harrison, B. D., and Mayo, M. A., 1980, Replication of RNA-1 of tomato black ring virus independently of RNA-2, *J. Gen. Virol.* **51**:317–326.

88a. Verkleij, F. N., De Vries, P., and Peters, D., 1982, Evidence that tomato spotted wilt virus RNA is a positive strand, *J. Gen. Virol.* **58**:329–338.

88b. Ie, T. S., 1982, A sap-transmissible, defective form of tomato spotted wilt virus, *J. Gen. Virol.* **59**:387–391.

89. Matthews, R. E. F., 1982, Classification and nomenclature of viruses, *Intervirology* **17**:142.

90. Vloten-Doting, L. van *et al.*, 1981, Tricorna: A proposed family of plant viruses with tripartite single-stranded RNA Genome, *Intervirology* **15**:198.

91a. Dougherty, W. G., and Kaesberg, P., 1981, Turnip crinkle vines RNA and its translation in rabbit reticulocyte and wheat embryo extracts, *Virology* **115**:45–56.

91b. Altenbach, S. B., and Howell, S. H., 1981, Identification of a satellite RNA associated with turnip crinkle virus, *Virology* **112**:25–33.

92a. Matthews, R. E. F., 1974, Some properties of turnip yellow mosaic virus nucleoproteins isolated in cesium chloride density gradients, *Virology* **60**:54–64.

92b. Keeling, J., and Matthews, R. E. F., 1982, Mechanism for release of RNA from turnip yellow mosaic virus at high pH, *Virology* **119**:214–218.

92c. Keeling, J., and Matthews, R. E. F., 1982, Mechanism of release of RNA from turnip yellow mosaic virus, *Virology* **119**:214.

93a. Koenig, R., 1979, CMI/AAB, No. 214.

93b. Matthews, R. E. F., 1982, Classification and nomenclature of viruses, *Intervirology* **17**:138.

93c. Paul, H. L., Gibbs, A., and Wittmann-Liebold, B., 1980, The relationship of certain tymoviruses assessed from the amino acid composition of their coat proteins, *Intervirology* **13**:99.

94. Chu, P. W. G., Francki, R. I. B., and Randles, J. W., 1983, Detection, isolation and characterization of high molecular weight double-stranded RNAs in plants infected with velvet tobacco mottle virus, *Virology* **126**:480–492.

95a. Kiefer, M. C., Owens, R. A., and Diener, T. V., 1983, Structural similarities between viroids and transposable genetic elements, *Proc. Nat. Acad. Sci. USA* **80**:6234.

95b. Dickson, E. A., 1981, A model for the involvement of viroids in RNA splicing, *Virology* **115:**216–221.

95c. Matthews, R. E. F., 1982, Classification and nomenclature of viruses, *Intervirology* **17:**179.

95d. Gross, H. J., Domdey, H., Lossow, C., Jank, P., Raba, M., Alberty, H., and Sänger, H., 1978, Nucleotide sequence and secondary structure of potato spindle tuber viroid, *Nature* **273:**203.

95e. Diener, T. O., and Smith, D. R., 1971, Potato spindle tuber viroid. VI. Monodisperse distribution after electrophoresis in 20% polyacrylamide gels, *Virology* **46:**498.

95f. Rackwitz, H.-R., Rohde, W., and Sänger, H. L., 1979, DNA-dependent RNA polymerase H of plant origin transcribes viroid RNA into full-length copies, *Nature* **291:**297.

96a. Gumpf, D. J., 1971, Purification and properties of soil-borne wheat mosaic virus, *Virology* **43:**588.

96b. Powell, C. A., 1976, The relationship between soil-borne wheat mosaic virus and tobacco mosaic virus, *Virology* **71:**483.

97. Nuss, D. L., and Peterson, A. J., 1981, Resolution and genome assignment of mRNA transcripts synthesized *in Vitro* by wound tumor virus, *Virology* **114:**399–404.

Phages of Prokaryotes (Bacteria and Cyanobacteria)[1]

---

[1] Partial list; many phages described incompletely or by only one laboratory are omitted.

Phage and host[1]	Family[2]
A (*Corynebacterium*)[(1)]: 50 nm diameter head and 83 nm tail	Syphoviridae
A (*Escherichia coli*): strain of ΦX 174	Microviridae
A-1 (L) (*Anabaena*): 60 nm diameter head, 14 × 83 nm tail contracting to 49 nm, with base plate 14 × 20 nm (cyanomyovirus)	Cyanophages
A1-1 (*Azospirillum*): 56 nm head, 250 nm tail with six spikes, DNA of 26×10^6 daltons with cohesive ends	Syphoviridae
A-2 (*Anabaena*): 63 nm head and 100 nm tail, DNA of 24×10^3 daltons (cyanomyovirus)	Cyanophages
A 4-L (*Anabaena*): 56 nm diameter head, 12 nm tail (cyanopodovirus)	Cyanophages
A 5/A 6 (*Alkaligenes faecalis*): 75 nm diameter head and 240 nm tail, serologically related to A25	Syphoviridae
A 6 (*Alkaligenes faecalis*): 90 nm diameter head and 110 nm tail	Myoviridae
All (*Azotobacter vinelandii*): 120 nm diameter head, 20 × 180 nm tail	Myoviridae
A 11/A 79 (*Alcaligenes faecalis*): tail lacking appendages	Myoviridae
A 12 (*Azotobacter vinelandii*): 56 nm diameter head	Podoviridae
A 13 (*Azotobacter vinelandii*): 60 nm diameter head, 10 × 170 nm tail	Myoviridae
A 14 (*Azotobacter vinelandii*): 120 nm diameter head, 20 × 180 nm tail, and 34% DNA of 165×10^6 daltons	Myoviridae
A 20 (*Asticacaulis*): long tail with appendages	Syphoviridae
A 21, A 22, A 23, A 24 (*Azotobacter vinelandii*), (serologically interrelated phages): 57 nm diameter head with short tail, 58% DNA	Podoviridae

[1] Host bacteria used less frequently than the common enterobacteria, e.g., *Escherichia, Salmonella, Pseudomonas,* and *Bacilli,* will be included in the alphabetical listing. Please note that many phages are identified by the Greek Φ before the letter or number. Thus search also under Φ. The Greek letters are placed after the Latin alphabetic listing, followed by number identifications (Arabic followed by Latin numerals).

[2] Names of genera, when proposed or accepted, are given in parentheses. Syphoviridae is the name approved by the ICTV for the group termed Styloviridae for a while and corresponds to Bradley's group B. A = Myoviridae, C = Podoviridae.

Phage and host	Family
A 25 (*Streptococcus*), (related to GT 234): 47 nm diameter head and tail with terminal knob, 8 × 185 nm, virulent transducing phage	Syphoviridae
A 31 (*Azotobacter vinelandii*): 61 nm diameter head, 26% DNA	Syphoviridae
A 41 (*Azotobacter vinelandii*): unstable particles, DNA of 46 × 10⁶ daltons	Podoviridae
A 64/A 62 (*Alcaligenes faecalis*): isometric with long, thin tail	Syphoviridae
A 422 (Brucella spp.)[2] identical to Tbilisi phage	
AC-1 (*Anacystis, Chroococcus*): 63 nm diameter head, 25 nm tail (cyanopodovirus)	Cyanophage
Ac 20 (*Asticacaulis*): isometric head	Syphoviridae
Acholeplasma phages: L 2, L 3, L 51, MVG, MVL	
Achromobacter (now classified as *Vibrio*) phages: α1, α2	
Acinetobacter phages: BP 1, P 78	
Actinophages: Phages of *Actinomyces, Corynebacterium, Arthrobacter, Mycobacterium, Streptomyces, Thermoactinomyces, Thermomonospora, Nocardia, Bifidobacterium*, etc.	
AE 2 (enterobacteria): 5 × 800 nm (43 S), (fd group of inoviruses)	Inoviridae
AG 8010: *see* Φ AG 8010	
Agrobacterium phages: B 179, LHII, LVI, P 0362, P 8149, BII BNV, PB 2A, PR, P 58, P S192, PSR, PT, R 4, Ψ, Ω	
Alcaligenes phages: A 5, A 6, A 11/A 79, A 64/A 62, 8764, 8893	
AN 10, AN 15 (cyanomyoviruses)	Cyanophages
AN 20, AN 22, AN 24 (cyanopodoviruses)	Cyanophages
Ancalomicrobium phages: EV, SP, Va	
AP 50 (*Anthrax*)[3]: 14% phospholipid containing phage, high strain specificity, buoyant density in sucrose 1.198 g/cm³, in CsCl, 1.301 (related to PR 3)	Tectiviridae
AR 1 (*Bacillus subtilis*): SP 8 group	Myoviridae

Phage and host	Family
AR 2, AR 3 (*Bacillus subtilis*): 100 nm diameter head, 200 nm tail with base plate with 100 nm fibers (SP50 group)	Myoviridae
AR 9 (*Bacillus subtilis*)[4]:contains uracil instead of thymine (PBS1 group)	Myoviridae
Arthrobacter phages: ΦAG 8010, NN	
AS-1 (M), (*Anacystis, Synechococcus* spp.): Cyanophage, type species of genus cyanomyovirus. The isometric head has a diameter of 90 nm and a tail of 23 × 244 nm, contracting to 93 nm, with short tail pins on a 40 nm base plate (750 S, buoyant density in CsCl 1.5 g/cm^3). The DNA is about 60 × 10^3 daltons, and there are about 30 proteins. (Cyanomyovirus)	Cyanophage
Asticacaulis phages: A 20, AC 20	
AT 298 (*Streptococcus*): unclassified	
Azotobacter phages: A 11–A 14, A 21–A 24, A 31, A 41	
B (*Corynebacterium diphtheriae*)	Syphoviridae
B 01 (B 02a, B 02b), (*Mycobacterium smegmatis*): similar or identical to Phlei	
B 01 (*Escherichia coli*): (subgroup I)	Leviviridae
B 1, B 3, B 33 (*Pseudomonas aeruginosa*): female-specific temperate phages, 50 nm diameter head, 135 nm tail with terminal knob (buoyant density in CsCl 1.473 g/cm^3), DNA of 25 × 10^6 daltons	Syphoviridae
B 6, B 7 (Enterobacteria)	probably Leviviridae
B 33 (*Pseudomonas aeruginosa*)[5]: related to B1	
B 179 (*Agrobacterium radiobacter*): unclassified	
BA 14 (*Escherichia coli*): related to T 7	
Bacillus megaterium phage (ATCC-specific): 42 × 10^6 dalton DNA	
Bacillus thuringiensis phages: GT 1, etc.; GV 1, etc.	
Bacterioides phages: ΦA 1: unclassified	

Phage and host	Family
B am 35 (*Bacillus anthracis*): unusual structure of isometric 63 nm head with spikes at vertices and at short tail (Figure 26)	Tectiviridae
Bdello vibrio phages: VL-1: unclassified	
Berne 6/29 (*Vibrio*)[6]: 66 × 74 nm head, short tail	Podoviridae
BF 23 (*Escherichia coli*)[7]: closely related to T 5	Syphoviridae
BG 3 (*Escherichia coli*): related to T5	Syphoviridae
Bk (*Brucella*)	Podoviridae
BL (*Corynebacterium diphtheriae*)	Syphoviridae
BLE (*Bacillus*)[8]: 43 × 120 nm head and 220 nm tail, 39 × 10⁶ dalton DNA	Syphoviridae
BM (*Bacillus*): 60 nm diameter head with 15 × 130 nm tail	
BP 1 (*Acinetobacter*): similar to P 78	Podoviridae
Brucella phages: many Podoviridae including A 422, M 51, S 708, Tbilisi (Tb), Bk, Wb, R	
Butyricum (*Mycobacterium butyricum*): 63 nm head with 213 nm tail with terminal knob (410 S), DNA of 116 × 10⁶ daltons	Syphoviridae
BZ 13 (*Escherichia coli*): (subgroup II)	Leviviridae
C (*Actinomyces*): *see* ΦC	
C (*Nocardium erythropolis*)[9]; 52 nm diameter head, 10 × 192 nm tail	
C 1 (*Micrococcus*)[10]: 45 nm diameter head	Podoviridae
C 1 (*Escherichia coli*): 90 nm diameter head, tail with pointed tip	Syphoviridae
C 5 (*Escherichia coli*), (inovirus)	Inoviridae
C 16 (*Escherichia coli*), (T-even group)	Myoviridae
C 22 (*Escherichia coli*), (inovirus)	Inoviridae
C 31 (*Streptococcus coelicolor*): temperate phage (inovirus)	Inoviridae
Caryophanum latum phages: ΦCL 29, CLV 29	
Caulobacter crescentes phages: many CB, cb, Cd, Cp, CP, CR phages, and Φ6	
CB 3 (*Pseudomonas aeruginosa*)[11]: unclassified	

Phage and host	Family

Cb K (*Caulobacter crescentes*): unclassified

Cb 5, Cb 8, Cb 13, etc. (*Caulobacter crescentes*): *see* ΦCb 5, ΦCb 8, ΦCb 13, etc.

Cd 1 (*Caulobacter crescentes*): 60 nm head, 11 nm   Podoviridae
tail, DNA of 28 × 10⁶ daltons

Cf (*Xanthomonas*)[12]: 1000-nm long, host-specific,   Inoviridae
not pathogenic, C/A = 2, with 2% of an uniden-
tified base (inovirus)

Chondrococcus phages: χ

Clostridian phages: F1, HM 2, HM 3, HM 7, k, M, S 2, 1, 80, α1, α2.

CORTICOVIRIDAE[13]: Icosahedral particles of about 60 nm diameter with
spikes at the vertices. The virion consists of a supercoiled DNA of 6
× 10³ daltons (12%), lipid, mostly phospholipid (13%), and four proteins (75%), (of 43, 32, 12, and 5 × 10³ daltons). Type species PM 2.

Corynebacterium phages: B, BL, β, γ, ω

Cp, CP: many *Caulobacter crescentus* phages, *see* ΦCp, ΦCP

CP 1 (*Bacillus cereus*): 90 nm diameter head and 20 × 160 nm tail

CP 1 (*Pseudomonas*): 55 nm head, 15 × 145 nm tail,   Myoviridae
30 × 10⁶ dalton DNA, chloroform-sensitive yet
no lipid

CP 1 (*Streptococcus pneumoniae*): Hexagonal head   Myoviridae
(45 × 60 nm) and 15 × 20 nm tail, head fibers,
and neck appendages. DNA of 19 S (12 × 10⁶ daltons), with covalently-linked protein of 28 × 10³
daltons.

CP 2, CP 18 (*Caulobacter crescentus*): *see* ΦCP 2, ΦCP 18

CP 3 (*Bacillus subtilis*): 66 nm diameter head and 13 × 276 nm tail

CP 51 (*Bacillus cereus*)[4]: transducing phage with   Myoviridae
90 nm diameter head, 20 × 160 nm tail, DNA
of 58 × 10⁶ daltons containing 5-hydroxymethyl-
uracil instead of thymine (SP8 group)

CP 53 (*Bacillus cereus*): transducing and lysogen-   Myoviridae
izing phage, head of 66 nm diameter, 13 × 276
nm tail, DNA of 17 × 10⁶ daltons

CR: many *Caulobacter crescentus* phages, *see* ΦCR, CR

CS-1 (*Bacillus megaterium*): 54 nm head, 10 × 200   Syphoviridae
nm tail

Phage and host	Family
CT 1 (*Rhizobium trifolii*): 60 nm diameter head, 23 × 118 nm tail	Myoviridae
CT 2 (*Rhizobium trifolii*)	Myoviridae
CT 3 (*Rhizobium trifolii*): 110 nm diameter head, 25 × 150 nm tail	Myoviridae
CT 4 (*Rhizobium trifolii*)[14]: 60 nm diameter head, 23 × 107 nm tail	Myoviridae
CT 5 (*Rhizobium trifolii*): 54 nm diameter head, 19 × 100 nm tail	Myoviridae
CT 6 (*Rhizobium trifolii*): 88 nm diameter head, 26 × 130 nm tail	Myoviridae

CYANOPHAGES[15]: Phages of blue-green algae (*Cyanobacteria*). All cyanophages contain a single molecule of double-stranded DNA and have typical phage heads and tails, resembling myo-, stylo-, or podoviridae. They are thus classified as genera under these family names. The type species of the *cyanomyoviruses* is AS-1. Its host range is *Anacystis* and *Synechococcus* species. Other members are N-1, A-1(L), and A-2 of filamentous cyanobacteria such as *Nostoc* (Anabaena).

The *cyanostyloviruses* (S 1, S-2L, and SM-2) of *Synechococcus*, and the last also of *Microcystis* have isometric heads of about 53 nm diameter, rigid (S-1) or flexible tails of 10 × 120–140 nm, DNA of about 25 × 10^6 daltons.

The *cyanopodoviruses* (type species LPP-1) of *Plectonema* and *Phormidium* have isometric heads of 50–70 nm diameter, tails of 15 × 15 nm, 27 × 10^6 dalton DNA and 10–14 proteins (548–820 S, buoyant density in CsCl 1.48 g/cm^3). Other members are LPP-2, SM-1 of *Synechococcus* and *Microcystis*, A4-L (of *Anabema*), and AC-1 of *Anacystis* and *Chroococcus*.

CYSTOVIRIDAE[16]; Isometric particles of about 60 nm diameter within a flexible lipid-rich (mostly phospholipid, 20% of particle weight) envelope (446 S, density in CsCL 1.27 g/cm^3). Three double-stranded RNAs of 2.3, 3.1, and 5.0 × 10^6 daltons and at least 11 proteins, including RNA polymerase. The host of the type species Φ6, a marine *Pseudomonas*, is infected via its pili, and lysed by the phage.

D (*Hydrogenomonas facilis*): similar to SH 133

D (*Escherichia coli*): temperate phage related to P 2	Myoviridae

D 1 (Blue-green algae): unclassified cyanophage

D 3 (*Pseudomonas aeruginosa*): 70 nm diameter head, 12 × 150 nm tail with six knob-like projections, 44 × 10^6 dalton DNA, temperate phage	Syphoviridae

Phage and host	Family

D 4 (*Mycobacterium* spp.): sensitive to chloroform, yet probably contains no lipid (*see* D 29)

D 5 (*Bacillus stearothermophilus*)   Myoviridae

D 6 (*Salmonella oranienburg*): unclassified

D 12 (*Escherichia coli*): (T-even group)   Myoviridae

D 28, D 29, D 29A, D 32 (*Mycobacterium* spp.): similar to D 4 (D 29: 65 nm diameter, variable tail length, three major, three minor proteins)

D 108 (*Escherichia coli*)[17]: general transducing   Myoviridae
phage, 65 nm diameter head, 22 × 173 nm tail

D 326 (*Escherichia coli*): related to Φ80

Dd II (*Shigella dysenteriae*): Head of 55 nm diameter   Podoviridae
and short tail (15 × 20 nm), (480 S, buoyant den-
sity in CsCl is 1.52 g/cm³). The phage contains
45% protein, 5% lipid, and 47% DNA (33 S) of 30
× 10⁶ daltons.

Dd VI (*Shigella dysenteriae*, also *Escherichia coli*):   Myoviridae
Elongated head (86 × 115 nm), and long contrac-
tile tail (18 × 130 nm), (859 S, buoyant density
in CsCl 1.48 g/cm³). The phage contains 48%
DNA (66 S, 143 × 10⁶ daltons) with glycosylated
5-hydroxymethylcytosine instead of cystosine (T-
even group).

Dd VII (*Shigella dysenteriae*, also *Escherichia*   Syphoviridae
coli)[18]: Head of 50 nm diameter and long noncontractile tail (4 × 135 nm), (530 S, buoyant
density in CsCl 1.52 g/ml). Possibly contains
5-methylcytosin and 7-methylguanine.

Diplophages (Streptococcus): DP 1–DP 4, W 3, W 8

DM 11, DM 21, DM 31 (*Levinea*): unclassified

DP-1, DP-2, DP-3 (*Streptococcus pneumoniae*):   Syphoviridae
Head of 67 nm diameter with envelope containing
lipid (8.5%), 150 nm tail without fibers (313 S,
buoyant density in CsCl 1.45 g/cm³), 67 × 10⁶
dalton DNA.

Dp-4 (*Streptococcus pneumoniae*)[19]: 60 nm diam-   Syphoviridae
eter head, 155 nm tail, buoyant density in CsCl
1.48 g/cm³, DNA of 37 × 10⁶ daltons, five proteins

Phage and host	Family
DS 6A (*Mycobacterium*): similar to D 4	
D Φ 3, D Φ 4, D Φ 5 (Enterobacteria)	Microviridae
e (*Rhizobium*) 82 nm head, 124 nm tail	Myoviridae
E 1 (*Escherichia coli*): octahedral head of 80 nm diameter, 100 nm tail with cross-striations and four fibers but no well-developed base plate	Myoviridae
E 79 (*Pseudomonas aeruginosa*): attaches to cell wall lipopolysaccharides, 38 S DNA	Myoviridae
EC (*Nocardia erythropolis*): 52 nm diameter head and 10 × 197 nm tail	
Ec 9 (Enterobacteria): 5 × 600 and 900 nm (possibly fd group of inoviruses)	Inoviridae
EJ (*Escherichia coli*): (subgroup II)	Leviviridae
ES 18 (*Salmonella*): general transducing phage, long tail	
E$_{sc}$-7-11 (*Escherichia coli*): long (bacilliform) head	possibly Podoviridae
EV (*Ancalomicrobium*): 57 nm diameter head and 5 × 148 nm tail lacking plate and fibers	Syphoviridae
F 1 (*Clostridium sporogenes*): 55 × 90 nm head, 10 × 123 nm flexible tail with cross-striations (301 S)	Syphoviridae
f 1 (Enterobacteria)[20]: possibly identical with fd	Inoviridae
F 1 (*Escherichia coli*)[6]: large octahedral head and thick tail with forked tip, type species	Syphoviridae
f 2 (*Escherichia coli*)[21]: type species (subgroup I)	Leviviridae
f 4 (*Escherichia coli*): (subgroup I)	Leviviridae
F 12 (*Escherichia coli*): (inovirus)	Inoviridae
F 116 (*Pseudomonas aeruginosa*): 65 nm diameter head, 80 nm tail, DNA of 38 × 10^6 daltons, attaches to pili	Syphoviridae
FA 5 (*Escherichia coli*): (subgroup I)	Leviviridae
FC 3-9 (*Klebsiella pneumoniae*): 110 nm looped tail	Myoviridae
fcam (*Escherichia coli*): (subgroup I)	Leviviridae
fd (Enterobacteria)[22]: type species of main group (inoviruses), (Figure 27)	Inoviridae

Phage and host	Family
FELS (*Escherichia coli*): recombinant of P 22, ES 18	
Ff (*Escherichia coli*); subgroup of Inoviridae	
FF 116 (*Pseudomonas* spp.): generalized transducing phage, attaching to pili, contains DNA of 39×10^6 daltons	
FH 5 (Enterobacteria): (subgroup IV, closely related to SP)	Leviviridae
FI (*Escherichia coli*): (subgroup IV)	Leviviridae
FJC (*Escherichia coli*): (subgroup IV)	Leviviridae
Flavobacterium phages: NCMB 384, 385	
fr 1 (*Escherichia coli*): (subgroup I)	Leviviridae
FV	Inoviridae
G (*Bacillus megaterium*): very large phage with head of 160 nm diameter and 455 nm tail (contracted 11×188 nm), 5×10^8 dalton DNA, type species.	Myoviridae
g 3, g 8, g 12, g 16, g 18 (*Bacillus subtilis*): unclassified	
G 4, G 6, G 13, G 14 (*Escherichia coli*)[23]: similar to ΦX 174 (G 6 serologically related, others not)	Microviridae
G 101 (*Pseudomonas aeruginosa*): temperate phage, 60×75 nm head, striated tail of 13×200 nm with terminal tassel, 37×10^6 dalton DNA	Syphoviridae
G III (Blue-green algae): very virulent, possibly identical to LPP-1 (cyanopodovirus)	Cyanophage
GA (*Escherichia coli*): (produces a relatively stable RNA replicase), (subgroup II)	Leviviridae
GA-1 (*Bacillus stearothermophilus, B. subtilis*)[6]: 45×60 nm head and 40 nm tail, collar with pins, morphology and properties similar to Φ29 (type species)	Podoviridae
GA/2 (*Bacillus subtilis*): defective phage, possibly identical to PBSX	
Gd, Ge, Gf (*Pseudomonas pseudoflora*)[24]: identical or very closely related to PBSX (pilus-dependent)	Syphoviridae
gh-1 (*Pseudomonas putida*): 50 nm diameter head, short wedge-shaped tail, buoyant density in CsCl 1.45 g/cm^3, 23×10^6 dalton DNA	Podoviridae

Phage and host	Family
GH 5, GH 8 (*Bacillus stearothermophilus*): thermophilic phage	
GR (*Escherichia coli*): (subgroup I)	Leviviridae
GSW (*Bacillus subtilis*): contains 5-hydroxmethyluracil instead of thymine	
GT-1, GT-2, GT-3, GT-4, GT-5 (*Bacillus thuringiensis*): temperate phages	
GT-6 (*Bacillus*)[6]: SP 50 group	Myoviridae
GT 234 (*Streptococcus*): transducing phage related to A 25	Syphoviridae
GV 3, GV 5 (*Bacillus*)[6]: elongated head (GA-1 group)	Podoviridae
GV 6 (*Bacillus*): (SP50 group)	Myoviridae
H (*Pasteurella pestis*): related to Φ I, Φ II	
H-1 (*Bacillus subtilis*): 100 nm head, 23 × 140 nm tail, the 83 × 10⁶ dalton DNA contains 5-hydroxymethyluracil instead of thymine	Myoviridae
Haemophilus phages: MP 1, N 3, S 2	
Halobacterium phage (*H. salinarum*): 57 nm head, 127 nm tail	Myoviridae
HF (*Hydrogenomonas facilis*): similar to SH 133	
HK 022 (*Escherichia coli*)[25]: temperate phage, 45 × 51 nm head, 106 nm flexible tail, DNA of 26 × 10⁶ daltons	Syphoviridae
HM 2 (*Clostridium saccharoperbutylacetonicum*)[6]: type species	Podoviridae
HM 3 (*Clostridium saccharoperbutylacetonicum*)[6]: type species	Myoviridae
HM 7 (*Clostridium saccharoperbutylacetonicum*)[6]: type species	Syphoviridae
HMT (Clostridium)[6]: (HM 2 group)	Podoviridae
HP 1 (*Haemophilus influenzae*), (mutants HP 1 c1 and c2): temperate phage of high transfection efficiency (10⁻³), related to S 2, 50 nm diameter head and 19 × 122 nm tail, DNA of 20 × 10⁶ daltons with cohesive ends	Myoviridae
HR (Enterobacteria): 6 × 800 nm (fd group of inoviruses)	Inoviridae

Phage and host	Family
hv-1 (*Vibrio*): 70 nm diameter head, 220 nm tail	Syphoviridae
Hydrogenomonas phages: D, HF, SH 133	
Hyphomicrobium phages: Hy Φ 30	
Hy Φ 30 (*Hyphomicrobium*), (294 S): (similar to T 7, PL 25), buoyant density in CsCl 1.508 g/cm³, DNA of 30 × 10⁶ daltons	Podoviridae
I 3 (*Mycobacterium smegmatis*): transducing phage, 80 nm diameter head with visible capsomers, and an 80 nm tail which contracts to 48 nm, contains 14% lipid	Myoviridae
I 10 (*Bacillus subtilis*): pseudotemperate, related to PBS 1	Myoviridae
ID 2 (*Escherichia coli*): (subgroup IV)	Leviviridae
If 1, If 2 (Enterobacteria): 5.5 × 1300 nm (inovirus)	Inoviridae
IKe (Enterobacteria): 6.6 × 1300 nm (inovirus)	Inoviridae

INOVIRIDAE[26]: A family composed of two rather dissimilar genera sharing three properties: the phage particles are long (threads and rods, respectively), contain single-stranded DNA, and are released by the host cell without lysis. The genera are termed *inovirus* and *plectrovirus*.

The *inoviruses* are helical threads (pitch 3.2 nm) of characteristic lengths for each species (6 × 750–2000 nm, density in CsCl 1.29 g/cm³). The single DNA molecule is circular and of corresponding lengths (1.7 × 10⁶ daltons for the fd subgroup). The DNA is generally high in T (33%). The coat protein consists of about 49 amino acids (molecular weight about 5000), and there are a few (2–4) molecules of a second protein of about 65 × 10³ daltons, the maturation protein. Several of these phages have been sequenced in terms of both DNA and protein. They carry about nine genes. Phages of this genus infect male strains of Enterobacteria including *Pseudomonas*, *Vibrio*, and *Xanthomonas*. Infection is generally at the tip of pili. Six recognized subgroups are (1) the Ff group of Enterobacteria and *Pseudomonas* (fd, f 1, AE 2, Ec 9, HR, M 13, 2 G/2, ZJ/2, δA, most about 800-, some 600- and 900-nm long), (2) If 1, If 2, and possibly Ike of Enterobacteria (about 1300-nm long), (3) Pf 1 and Pf 2 of *Pseudomonas* (about 1900-nm long), (4) Cf, Xf, Xf 2 of *Xanthomonas* (980-nm long), and (5) v 6 of *Vibrio*.

The other genus is the *plectroviruses*. These are straight rods possibly icosahedral, usually 14 × 84 nm, density in CsCl 1.37 g/cm³, consisting of a circular DNA molecule of 1.5 × 10⁶ daltons and four proteins of 19–70 × 10³ daltons. Their host is *Acholeplasma*. Members are MVL 51, (the type species), MV-L 1, MVG 51, P 3 c/r, 10 tur, etc.; a possible member is SV-C 1 of *Spiroplasma* (~13 × 250 nm).

Phage and host	Family
J 1 (*Lactobacillus casei*): 55 nm diameter head, flexible 10 × 290 nm tail, buoyant density in CsCl 1.49 g/cm³, DNA of 24 × 10⁶ daltons with cohesive ends (though phage is virulent)	Syphoviridae
Jersey (Enterobacteria): isometric	Syphoviridae
JP 34 (Enterobacteria): (subgroup II)	Leviviridae
JP 501 (Enterobacteria): (subgroup I)	Leviviridae
K (*Clostridium sporogenes*)[28]: (F 1 group)	Syphoviridae
K (*Staphylococcus aureus*): 70 nm diameter head, 15 × 210 nm tail with complex basal appendages (buoyant density in CsCl 1.479 g/cm³), DNA of 33 × 10⁶ daltons	Myoviridae
K 1 (*Streptococcus coelicolor*)	Syphoviridae
K 19 (Enterobacteria): isometric	Myoviridae
KB 1 (smooth *Salmonella* strains)[29]: 60 nm diameter head, 15 × 25 nm tail with complex base plate	Podoviridae
KJ (*Escherichia coli*): (subgroup II)	Leviviridae
Klebsiella phages: FC 3–9, Mp.	
KT (*Clostridium*): (HM3 group)	Myoviridae
KU 1 (*Escherichia coli*): (subgroup II)	Leviviridae
KZ (*Pseudomonas*): 120 nm head and 180 nm tail, DNA of over 200 × 10⁶ daltons	Myoviridae
L (*Salmonella typhimurium*): related to P 22	Podoviridae
L 1, L 2, L 3, L 51 (mycoplasma viruses): *see* MVL 1, etc.	
L-17 (Enterobacteria): closely related or identical to PRD 1, PR 3, etc.	
Lactobacillus phages: J 1, PL 1	
Lambda: *see* λ	
Leo (actinophage)	Syphoviridae
Levinea phages: DM 11, DM 21, DM 31: unclassified	

LEVIVIRIDAE[30]: The family name for the classical group of RNA phages. Members of the only classified genus (*levivirus*) have tentatively been divided into two physicochemically different groups (A and B), each consisting of two serologically differing subgroups (A = I, II; B = III,

Phage and host	Family

LEVIVIRIDAE[30]:   (continued)

IV) with some phages showing intermediate properties. They are all icosahedral, about 23 nm in diameter, 80 S, and have a buoyant density in CsCl of 1.44–1.47 g/cm^3. They consist largely of 180 molecules (32 capsomers, T = 3) of a capsid protein of 13–14 × 10^3 daltons (Group A) and about 17 × 10^3 daltons in Group B, all lacking histidine and, at times, other amino acids. The virions also carry one molecule of a "maturation" or "adsorption" (A) protein (about 45 and 48 × 10^3 daltons in Groups A and B, respectively) that serves the role of phage tails in being required for the typical attachment of the phage to the side of the pili of generally male enterobacteria (*Escherichia coli*, etc.). (This is in contrast to the Inoviridae which attach to the tip of the pilus). Group B has an additional virion protein (A 1) of as yet unknown function (about 39 × 10^3 daltons). (Fig. 27).

The RNA of the Leviviridae is a single plus-strand molecule of about 1.21 × 10^6 daltons in Group A (subgroups I and II) and 1.39 × 10^6 daltons in Group B (subgroups III and IV), since the latter carry the additional information for the A 1 protein. Besides the virion proteins, the Leviviridae carry a gene for a peptide chain of about 57 × 10^3 daltons which together with three host proteins forms the RNA polymerase necessary for the replication of these viruses. In addition, a lysis protein was detected in group A (8 × 10^3 daltons), overlapping the end of the coat gene, the intergenic sequence, and the beginning of the polymerase gene. Such a protein has not yet been identified in group B phages. In the latter, the coat and A 1 genes are coded in the same phase, the proteins starting with identical N-termini and A 1 only continuing by read-through of the weak coat gene termination signal. Group B, possibly because of its longer RNA, is more UV-sensitive than Group A. Several of these phages have been sequenced.

All Leviviridae lyse the host cell within less than an hour, releasing several thousand progeny. Members of Group I are MS 2 (the type species), f 2, f 4, fr, R 17, JP 501, FR 1, Zr, BO 1; of Group II, GA, SD, TH 1, BZ 13, KU 1, JP 34, KJ; of Group III, Qβ, VK, ST, TW 18; and of Group IV, SP, FI, TW 19, TW 38, MX 1, and ID 2—all attacking enterobacteria. Among other not yet classified Leviviridae are several *Caulobacter* phages (many of the Φ CB, Φ CP, and Φ Cr groups) and PPR 1, PP 7, and 75 of *Pseudomonas*. Certain *Bdellovibrio* phages may also be Leviviridae.

Phage and host	Family
LH II (L II) BNV (6-1, and 6-2), and LH I (L II) BV (7-1 and 7-2), (*Agrobacterium tumefaciens*): 72 nm diameter head and 235 nm flexuous tail with six appendages near end (591 S), buoyant density in CsCl 1.51 g/cm^3	Syphoviridae

LL 55: *see* Φ LL 55

Phage and host	Family
LP 51, LP 52 (*Bacillus licheniformis*)[8]: temperate interrelated phages (*see* Φ 1)	Syphoviridae
LPP-1, LPP-2, (LPP-3) (*Lyngbya, Plectonema, Phormidium*): 58 nm head, 20 × 15 nm tail (~500 S, buoyant density in CsCl 1.48 g/cm³), (cyanopodovirus)	Cyanophage
LT 2 (*Salmonella typhimurium*): similar to L	
LV 1 (*Agrobacterium tumefaciens*), (very similar to or identical with ω, PB 2A, PS 8): Particles with hexagonal head of 70 × 80 nm diameter with flexuous tail of 18 × 280 nm, buoyant density in CsCl 1.505g/cm³, DNA of 34 × 10⁶ daltons. The main proteins have molecular weights of 48 × 10³ (50%), 30 × 10³ (30%), 16 × 10³ (16%), and 69 × 10³ (4%).	Syphoviridae
M (*Clostridium sporogenes*): large and small particles (M1 and Ms) with 83 and 48 nm diameter head, 10 × 337 and 13 × 177 nm tail	
m (*Rhizobium*): 110 nm head, 144 nm tail	Myoviridae
M 1 (*Streptococcus*): elongated head	Syphoviridae
M₁ (*Thermoactinomyces*): 50 × 62 nm head and 90 nm tail	Syphoviridae
M 2 (*Bacillus subtilis*): (related to GA-1 and β3) carries a protein at ends of DNA strands	Podoviridae
M 6 (Enterobacteria): (inovirus)	Inoviridae
M 12 (*Escherichia coli*): (subgroup I)	Leviviridae
M 13 (Enterobacteria): closely related to fd (inovirus)[26]	Inoviridae
M 20 (Enterobacteria)	Microviridae
M 51 (Brucella spp.): similar, but not serologically related to Tbilisi phage	
MB 78 (*Salmonella typhimurium*)[31]: 60 nm diameter head, 95 nm tail with large basal knob	Syphoviridae
Methanomonas phages: MP 1–MP 3	
MG 40 (*Salmonella typhimurium*): similar to L	Podoviridae
Micrococcus phages: C 1, N 1–N 8, W, X	

Phage and host	Family

MICROVIRIDAE[32]: Icosahedral virions of 27 nm diameter with bulky spikes on the 12 vertices (particle weight 6.7×10^6, 114 S, density in CsCl 1.40 g/cm³). The particles contain one molecule of single-stranded circular DNA, usually high in T (33%), and of 1.7×10^6 daltons. The four proteins are the 60 molecules of the capsid protein of about 48×10^3 daltons, 12 pentameters of 19×10^3 and 5×10^3 dalton proteins forming the spikes, and a protein of 36×10^3 daltons forming their tips. Several microviruses have been sequenced. There are nine genes, several of them overlapping and transcribed in different phases. These phages adsorb by their spikes to the cell wall of their hosts, enterobacteria, and are released by cell lysis. The type species is Φ X 174, and other members are d Φ 3, d Φ 4, d Φ 5, G 4, G 6, G 13, G 14, M 20, SA-1, S 13, St-1, U 3, WA/1, Wf/1, α 3, α 10, δ 1, ζ 3, η 8, o 6, Φ A, Φ R, 1 Φ 1, 1 Φ 3, 1 Φ 7, 1 Φ 9. Three groups have also been defined: Group A (member α 3) infects *E. coli* strains B and C, Group B (Φ X 174, Φ R, S 13) infect strain C but not B and U 12 (they are serologically related to Groups A and C), Group C (member St-1) infects *Escherichia coli* K 12, not strains B and C

Mini-phages: satellite phages, such as mini-M 13 (half length), (inovirus)	Inoviridae
mor 1 (Bacillus): 43 × 88 nm head, 146 nm tail, and 28 × 10⁶ dalton DNA, type species (Figure 28)	Syphoviridae
Mp (*Aerobacter aerogenes, Escherichia coli*, and *Klebsiella pneumoniae*, not *Salmonella, Proteus, Serratia* spp.): hexagonal 62 nm head, flexuous tail of 6 × 165 nm, DNA of 23 × 10⁶ daltons	Syphoviridae
MP 1, MP 2, MP 3 (*Methanomonas methylovora*): 100 nm diameter head and 30 × 100 nm tail	
MP 7 (*Bacillus megaterium* QMB 1551): 60 × 70 nm head, thin 170 nm tail, DNA of 42 × 10⁶ daltons	Syphoviridae
MP 13 (*Bacillus megaterium*): odd-shaped head, 200 nm tail; note complex tail spikes (Figure 29)	Myoviridae
MP 15 (*Bacillus megaterium*): round head, 300 nm tail; note small spheres attached to tail (Figure 30)	Syphoviridae
MS 2 (*Escherichia coli*): type species (subgroup I)	Leviviridae
MSP 8 (*Streptomyces griseus*): 56 × 71 nm head and 158 nm tail (Fig. 27)	Syphoviridae
MU(1): *see* μ(1)	

Phage and host	Family
MV Br 1 (*Acholeplasma*): 72 nm diameter head, long tail	Probably Syphoviridae
MVG 51 (*Acholeplasma laidlawii*): (plectrovirus)	Inoviridae
MV Lg-pS 2L (*Acholeplasma laidlawii*)	Plasmaviridae
MVL 1 (*Acholeplasma laidlawii*): not lytic nor cytocidal (plectrovirus)	Inoviridae
MVL 2 (*Acholeplasma laidlawii*)[33]: not lytic nor cytocidal (type species)	Plasmaviridae
MVL 3 (*Acholeplasma laidlawii*): not lytic but cytocidal, 60 nm head, a thin collar with attached fibers, a 10 × 20 nm tail, 26×10^6 dalton DNA	Podoviridae
MVL-51 (*Acholeplasma laidlawii*): not lytic nor cytocidal (type species of plectrovirus)	Inoviridae
MX (*Escherichia coli*): intermediate between subgroup III and IV (group B)	Leviviridae
MX 1 (*Myxococcus xanthus*): 90 nm diameter head, thin collar, 100 nm tail with base plate and pins (1145 S), buoyant density in CsCl 1.531 g/cm³, 23 proteins	Myoviridae
MX 4, 41, 43 (*Myxococcus xanthus*): related generalized transducing phages with 67 nm head, 22 × 120 nm tail, DNA of 39×10^6 dalton, terminally redundant, but apparently lacking cohesive ends	Myoviridae
MX 8, 81, 82 (*Myxococcus xanthus*): related generalized transducing phages, 60 nm diameter head	Podoviridae
MX 9 (*Myxococcus xanthus*): 60 nm diameter head, dubious tail	Podoviridae
MY (*Escherichia coli*): (subgroup I)	Leviviridae
Mycobacterium phages: *Butyricum, Phlei,* B 01, D 4, I 3, R 1, Leo	
Mycoplasmaviruses[33]: Type 1: (plectrovirus) not cytolytic,	Inoviridae
type 2: not cytolytic,	Plasmaviridae
type 3: not yet classified, cytolytic.	Podoviridae

MYOVIRIDAE (corresponding to Bradley's group A)[34]: Large phages with medium to long complex contractile tails. The T-even phage group of

Phage and host	Family

MYOVIRIDAE (*continued*)

enterobacteria represents the best known genus (*see* under *T-even phage* group). Others have not yet been formally classified. Among these are phages with either elongated heads (9266 and 16–19 of enterobacteria and 108/106 of *Thermomonospera*), or isometric heads (K19, o1, P1, P2, ViI, and 121 of enterobacteria, PIIBNV6 of *Agrobacterium*, A 6 of *Alcaligenes*, G, PBS1, SP3, SP8, SP-15, and SP50 of *Bacillus*, HM3 of *Clostridium*, J3 of *Mycobacterium*, CP1, PB-1, PP8, K2, W-16, and 125 of *Pseudomonas*, CT4, e, m, and WT1 of *Rhizobium*, Twort of *Staphylococcus*, RZb of *Streptococcus*, and XP5 of *Xanthomonas*).

Myxococcus phages: MX 1, MX 4 (41, 43), MX 8 (81, 82)

N 1 (*Micrococcus lysodeicticus*): 64 nm diameter head, 227 nm tail, DNA of 32 × 10^6 daltons, related to 186, but virulent, though DNA contains cohesive ends	Syphoviridae
N-1 (*Nostoc* = *Anabaena*): isometric 61 nm head, 100 nm tail with rigid spikes, and two flexible fibers at neck (539 S, buoyant density in CsCl 1.500 g/cm^3), DNA of 43 × 10^6 daltons, 19 proteins (cyanomyovirus)	Cyanophage
N 3 (*Haemophilus influenzae*): 60 nm head and 200 nm tail lacking terminal fiber(s), 26 × 10^6 dalton DNA with cohesive ends	Syphoviridae
N 4 (*Escherichia coli* K12): 70 nm diameter head, with complex base plate, (436 S), DNA of 40 × 10^6 daltons, ten proteins	Podoviridae
N 5, N 6 (*Mycrococcus lysodeicticus*): similar to λ	
N 1–N 8 (*Micrococcus luteus*): N 1–N 4, N 7, N 8, similar morphology, different from N 5 and N6; the DNA has cohesive ends	Syphoviridae
N 8 (Enterobacteria)	Microviridae
N 17 (Bacillus)[8]: SP 50 group	Myoviridae
N 17 (*Shigella flexneri*)[35]: contains 5-hydroxycytosin instead of cytosine	
NC MB 384, 385 (*Flavobacterium cytophaga* spp.)	Syphoviridae
NF (*Bacillus subtilis*)[8]: Member of Φ 29 group	
NH, NM (*Escherichia coli*): (subgroup III)	Leviviridae

Phage and host	Family
NM-1 (*Rhizobium*): isometric	Syphoviridae
NN (*Arthrobacter*): 75 nm diameter head and 220 nm tail	Syphoviridae
NN (*Bifidobacterium*): 60 nm diameter head and 215 nm tail	Syphoviridae
Nocardia phages: Φ C, EC, R1	
NS 11 (*Bacillus acidocaldarius*): acido-thermophile phage, contains 12% lipid, 14% DNA, virion of 75 nm diameter with spikes at vertices	Corticoviridae
NT 1 (*Rhizobium trifolii*): 62 nm diameter head, 12 × 140 nm tail	Syphoviridae
nt-1 (*Vibrio*): 72 × 137 nm diameter head, 113 nm tail	Myoviridae
NT 2 (*Rhizobium trifolii*): 60 nm diameter head, 12 × 170 nm tail	Syphoviridae
NT 3 (*Rhizobium trifolii*): 57 nm diameter head, 15 × 130 nm tail	Syphoviridae
NT 4 (*Rhizobium trifolii*): 60 nm diameter head, 9 × 100 nm tail	Syphoviridae
O 1 (Enterobacteriae): isometric	Myoviridae
O 3 clr (Plectrovirus)	Inoviridae
O6N-22P (*Vibrio*): 45 × 75 nm diameter head, 65 nm tail	Myoviridae
O 11 clr (Plectrovirus)	Inoviridae
Omega: *see* ω and Ω	
Omicron: *see* o	
OXN-52P (*Vibrio*): 70 nm diameter head, 120 nm tail	Syphoviridae
OXN-100P (*Vibrio*): 65 nm diameter head, 20 nm tail	Podoviridae
P 0362 (*Agrobacterium tumefaciens*): defective phage, head of 65 nm diameter and straight 130 nm tail, DNA of 25×10^6 daltons, buoyant density in CsCl 1.458 g/ml, lower T_m and different protein than the LV 1 group	Syphoviridae

Phage and host	Family

P 1 (*Salmonella, Escherichia coli*): Generalized transducing phage of 65 nm diameter with 20×200 nm tail. DNA of 60×10^6 daltons. The viral DNA is not integrated into the host chromosome, but resides in the cell as a plasmid. Specific restriction and modification systems were first studied in this phage. The DNA is circularly permuted; it contains 0.3% N^6 methyladenine and half as much 5-methylcytosine. Pseudovirions are often present

    Myoviridae

P lc3, mutant of P 1

Plkc (*Escherichia coli*): particle of 90 nm diameter with 20×220 nm tail, buoyant density 1.47 g/cm^3 in CsCl

    Myoviridae

P 2 (*Salmonella* and *Escherichia coli*)[36,37]: Temperate phage. Polyhedral, probably icosahedral particles of 58 nm diameter with cylindrical contractile tail of 17×135 nm, with thin tail fibers, 40–50 nm long.

    Myoviridae

The DNA (38%) has a molecular weight of 22×10^6. Its length is 13.2 μm. It is linear and nonpermuted, with cohesive ends consisting of 19 single-stranded nucleotides. It coheres with the DNA phages 186 and 299, even though their cohesive sequences differ slightly (by one nucleotide). The cohesive sequences of P 2 and P 4 are identical, but quite different from those of unrelated λ.

The head consists to 90% of the major capsid protein and contains at least six minor components. The major protein is cleaved from 44,000 to 36,000 daltons in the course of phage maturation, coincident with release of two of the minor proteins derived from the same gene (N). There are also at least four tail proteins (Figure 31).

P 3 (*Salmonella potsdam*): Hexagonal head of 48×55 nm with 12×118 nm tail, the sheath of which contracts to 42 nm, exposing the 5.5 nm core. Six short fibers at the end of the core are evident. The phage is very heat resistant.

    Myoviridae

P 4 (*Escherichia coli*): A defective satellite of P 2. The isometric particles have a diameter of 40 nm and a contractile tail very similar if not identical to that of P 2 (17×135 nm), with thin 45-nm

    Myoviridae

Phage and host	Family

P 4 (*Escherichia coli*): (*continued*)
long fibers. The DNA has the same cohesive ends as that of P 2, but is only one-third as long, 3.9 μm, molecular weight about 7×10^6 (the capacity of the head is also one-third of that of P 2). There is no detectable homology (<1%) between the DNA of P 2 and P 4. The proteins of P 4 appear to be the same as those of P 2 (Figure 31).

P 4 (*Salmonella potsdam*)[36,37]: hexagonal head of 48 × 55 nm with short tail (9 × 15 nm) — Podoviridae

P 9a (*Salmonella potsdam*): very similar to P 3

P 9c (*Salmonella potsdam*): very similar to P4

P 10 (*Salmonella potsdam*), (serologically related to P 3, P 9a): Hexagonal head of 48 × 55 nm with 12 × 95 nm tail, which contracts to 38 nm, exposing 6 nm core. Base plate carrying six long straight fibers. Very heat sensitive. — Myoviridae

P 11 (*Staphylococcus aureus*): general transducing phage — Syphoviridae

P 11-M 15 (*Staphylococcus aureus*): A virulent mutant of the temperate phage P 11. Particles of 50–60 nm diameter (445 S) with 150 nm flexuous noncontractile tail with terminal knob. DNA of 33×10^6 daltons, lacking cohesive ends or nicks. — Syphoviridae

P 22 (*Salmonella*, smooth strain)[38]: Temperate general transducing phage (which can integrate at several sites, not as many as μl). Isometric 60 nm diameter particles with short six-pin tail assembly. DNA is 26×10^6 daltons. It is, like those of P1 and the T-even phages, circularly permuted and terminally redundant. The DNA contains small amounts of N^6-methyladenine, which vary in different hosts. — Podoviridae

The main head protein (55×10^3 daltons) forms proheads in conjunction with a scaffolding protein (42×10^3 daltons) which leaves the maturing phage. Minor proteins of the head are 94, 67, 50, and 18×10^3 daltons. Those of the short tail assembly and pins are 23 and 76×10^3 daltons. The full heads, proheads, and empty heads are 500 S, 240 S, and 170 S. Pseudovirions are often present.

Phage and host	Family
P 42 D (Staphylococcus B): 55 nm diameter head with 230 nm tails	possibly Syphoviridae
P 52 A (Staphylococcus B): 50 nm diameter head with 150 nm tails	possibly Syphoviridae
P 78 (*Acinetobacter*, strain 78-specific): generalized transducing phage with 50 nm diameter head, 13 nm tail, buoyant density in CsCl 1.524 g/cm^3	Podoviridae
P 932a (*Pasteurella pestis*)	
P 8149 (*Agrobacterium radiobacter*): defective phage not related to LV 1 group of crowngall-associated factors. Buoyant density in CsCl is 1.510 g/cm^3. Bipyramidal particles with 40 nm diameter head and short tail, DNA of 10 × 10^6 daltons	Podoviridae
P II BNV 6 (*Agrobacterium radiobacter*). isometric	Myoviridae
P II BNV 6-C (*Agrobacterium radiobacter*): isometric	Podoviridae
P-a-1 (*Streptomyces*): isometric	Syphoviridae
PA 2 (Enterobacteria): (λ phage group)	Syphoviridae
Pasteurella phages: H, Φ I, Φ II, P 932a	
PB-1 (*Pseudomonas aeruginosa*): 75 nm diameter head with 150 nm tail, no base plate, and four 60 nm tail fibers folded back against the sheath; the DNA is 25 μm long	Myoviridae
PB 2 (*Pseudomonas syringae*): particles with elongated head (10 × 70 nm), and noncontractile tail of 175 nm; member of XP 12 group	Syphoviridae
PB 2A (*Agrobacterium tumefaciens*): related or identical with ω, PS 8, and LV 1, buoyant density in CsCl is 1.505 g/cm^3	Syphoviridae

PB 6, 7, 8, 9, 10, 29, 84, 1197 (*Pseudomonas aeruginosa*): contain DNA of 31 S, 23 × 10^6 mol. wt.

PBA 12 (*Bacillus subtilis*): bacilliform head (35 × 100 nm), 200 nm tail

PBLA, PBLB (*Bacillus licheniformis*): similar to PBSX (orphan pseudovirions)

| PBP 1 (*Bacillus pumilus*): transducing phage infecting only flagellated strains of bacteria, head of 84 nm diameter and 260 nm tail with flexuous fibers, DNA of 35 × 10^6 daltons, type species | Syphoviridae |

Phage and host	Family
PBS1 (*Bacillus subtilis*)[4]: Large pseudotemperate phage, 113 nm diameter head, 220 nm tail, 180 × 10^6 dalton DNA containing uracil instead of thymine, infects only flagellated strains, type species	Myoviridae
PBX 2 (*Bacillus subtilis*)[4]: related and similar to PBS 1	
PBS 2c (*Bacillus subtilis*): 90 nm diameter head, 10 × 280 nm tail with neck plate	Syphoviridae
PBSX (*Bacillus subtilis*), (also called PBSH and GA/2)[39]: A defective phage carrying only host DNA. The hexagonal particles have a diameter of 41 nm, a contractile tail of 18 × 196 nm with 70 nm tail fibers, 160 S, 1.375 g/cm³ buoyant density in CsCl. The DNA (of the host) is linear and of 9–12 × 10^6 daltons.	Myoviridae
PBV 1, PBV 3 (*Penicillium brevicompactum*, infectious for *E. coli*): 45 nm diameter head with long tail, buoyant density in CsCl is 1.48 g/cm³ for PBV 1 and 1.51 for PBV 3, contains linear DNA	Syphoviridae
PBV 2 (*Penicillium brevicompactum*, infectious for *E. coli*): particles of 53 nm diameter with a short tail, buoyant density in CsCl is 1.45 g/cm³, 610 S, linear DNA of 25.2 × 10^6 daltons	Podoviridae
Pc (*Pseudomonas aeruginosa*): 60 nm diameter head, 165 nm tail with terminal knob	Syphoviridae
PE 69 (*Escherichia coli*): (inovirus)	Inoviridae
PF 1 (*Clostridium*)[6]: (HM 2 group)	Podoviridae
Pf 1, Pf 2 (Enterobacteria): 1900 nm (inovirus)	Inoviridae
Pf 3 (*Pseudomonas* and *enterobacteria*): 760 nm (inovirus)	Inoviridae
Phlei (*Mycobacterium*): 63 nm diameter head, 158 nm tail (490 S, buoyant density in CsCl 1.51 g/cm³), DNA of 123 × 10^6 daltons	Syphoviridae
PK (*Enterobacteria*): mutant of P 2	
PK 1 (*Streptococcus mutans*): linear and circular DNA of 28 × 10^6 daltons	
pKC (*Bacterium subtilis*): 63 S DNA	
PL 1 (*Lactobacillus casei*): 50 nm diameter head, 11 × 275 nm tail with 50 nm terminal fibers, DNA of 25 × 10^6 daltons with cohesive ends	Syphoviridae

Phage and host	Family

PL 25 (P1 26, PL 27) (*Proteus morganii*, Providence   Podoviridae
strains): particles of 485 S containing 26 × 10^6
dalton DNA, 12.6 μm long (similar to P 22 of *Salmonella*)

PLASMAVIRIDAE[40a] (also termed mycoplasma virus type 2): Widely varying
pleomorphic enveloped particles (50–120 nm diameter) containing supercoiled circular DNA of 7.6 × 10^6 daltons, more than eight proteins,
and a lipid-containing envelope similar in composition to the host's
(Acholeplasma). The progeny virus is released by budding without cytopathology. The type species is phage MVL 2, and MVL 3, 1307, MV
Lg-pS2L, v 1, v 2, v 4, v 5, and v 7 are members

PLECTROVIRUSES[41]: genus of Inoviridae (also termed Mycoplasma virus
type 1) (see Inoviridae)

PLS 1: *see* Φ PLS 1

PLT 22 (*Salmonella typhimurium*): temperate phage

PM 2 (Marine *Pseudomonas* BAL-31)[42]: type spe-   Corticoviridae
cies

Pneumococcus phages: ω 1–ω 3, ω 7–ω 9

PO 2, PO 4 (*Pseudomonas aeruginosa*): 58 nm di-   Syphoviridae
ameter head and 186 nm tail with bar-shaped
basal structure, phages attach to pili

PODOVIRIDAE (corresponding to Bradley's group C)[43]: Phages of female
enterobacteria that are characterized by a very short (less than 20 nm)
and noncontractile tail. The type species of a genus (group) is T 7, and
T 3 is very similar to it. The phages are virulent and utilize the host's
degraded DNA in their replication. Members of this group are H, PTB,
R, Y, W 31, and Φ I and Φ II. Members of other potential genera are
the isometric N 4, P 22, sd, Ω 8, 7480b of enterobacteria, P II BNV 6-
C of *Agrobacterium*, Tb of *Brucella*, HM 2 of *Chlostridium*, C 1 of
Micrococcus, gh-1 of *Pseudomonas*, Φ 2042 of *Rhizobium*, Φ 17 of
Streptomyces, and 114 of *Thermomonospera*. Podoviridae with elon-
gated heads are 7–11 of enterobacteria, GA-1 and Φ 29 of *Bacillus*, and
182 of *Streptococcus*.

PP 1 (*Pseudomonas aeruginosa*): probably idential to PB 1

PP 4 (*Pseudomonas aeruginosa*): very similar or identical to PO 2

PP 7 (*Pseudomonas aeruginosa*, Enterobacteria):   Leviviridae
(subgroup I, similar to R 17)

PP 8 (*Pseudomonas aeruginosa*): 100 nm diameter   Myoviridae
head and 190 nm tail

Phage and host	Family
PPR 1 (*Pseudomonas aeruginosa*)	Leviviridae
PR 3, PR 4, PR 5, PR 772 (*Pseudomonas* and Enterobacteria)[44]: closely related if not identical also with PRD 1, pilus-specific phages, head of 53–65 nm diameter and tail of variable length and possibly not phage-specific, 11×10^6 dalton DNA with cohesive ends	Tectiviridae
PR 590a (*Agrobacterium radiobacter*)	Podoviridae
PR 772 (*Proteus mirabilis*): *see* PR 4	
PR 1001 (*Agrobacterium radiobacter*)	Podoviridae
PRD 1 (*Pseudomonas* and other gram-negative bacteria): The DNA carries a 5′ terminal protein, type species	Tectiviridae
PRM 1 (*Rhizobium meliloti*): unclassified	
PRR 1 (*Pseudomonas*, Enterobacteria)	Leviviridae
PS 4 (*Pseudomonas syringae*): 64 nm head and 170 nm tail	Syphoviridae
PS 8 (*Agrobacterium tumefaciens*): possibly identical to LV 1, PB 2A	
PS 192 (*Agrobacterium tumefaciens*): similar to PR 1001	Podoviridae
Psp 231a (*Pseudomonas phaseolicola*): 55 nm diameter head, buoyant density in CsCl 1.48 g/cm³ (407 S), 28×10^6 dalton DNA	Podoviridae
PsR 1012 (*Agrobacterium radiobacter*): similar to PR 1001	Podoviridae
PST (*Escherichia coli*): (T-even group)	Myoviridae
PT 11 (*Agrobacterium*): 55×110 nm head, 130 nm tail	Syphoviridae
PTB (*Yersinia pestis*)	possibly Podoviridae
PX (*Pseudomonas* phages related to CB 3: PX 1, 4, 10, 12, 14 psychrophilic, PX 2, 3, 5, 7 mesophilic)	
Qβ (*Escherichia coli*)[45a,b]: (subgroup III)	Leviviridae
R (*Brucella*)	Podoviridae
R (*Hydrogenomonas facilis*): similar to SH 133	Podoviridae

Phage and Host	Family
R 1 (*Mycobacterium butyricum*): bacilliform head of 50 × 97 nm, tail of variable length (~220 μm) with terminal knob and fibers, buoyant density in CsCl 1.472 g/cm^3, DNA of 25 × 10^6 daltons, 14% lipid	Syphoviridae
R 1 (*Nocardia*): 75 nm diameter head and 330 nm tail	Syphoviridae
R 1 (*Streptomyces coelicolor*): isometric	Syphoviridae
R$_2$ (*Streptomyces coelicolor*): 56 × 100 nm head and 170 nm tail	Syphoviridae
R 4 (*Agrobacterium tumefaciens*): (Related to PS 8, PB 2A, LV 1, and ω): Polyhedral head of 65 nm diameter with flexuous tail of 10 × 210 nm. Buoyant density in CsCl 1.51 g/ml. DNA of 34 S, 30 × 10^6 mol. wt. The phage contains four major proteins of 72, 45, 28, and 14.5 × 10^3 mol. wt.	Syphoviridae
R 17 (*Escherichia coli*)[46]: (subgroup I)	Leviviridae
R 23 (*Escherichia coli*): (subgroup I)	Leviviridae
Rhi φ I (*Rhizobium melilotis*), complex tail with spikes (Figure 32)	probably Myoviridae
Rhi φ L 9 (*Rhizobium melilotis*), complex tail with spikes (Figure 33)	probably Myoviridae

Rhizobium phages: CM1, CM2 CT 1–CT 4, m, MM1, NM1, NM2, NM3, NT 1–NT 4, PRM 1, st 1, WT 1, WT 2, 7-7-7, 16-6-12, 16-12-1, 317, Φ 2042, etc.

Rho: *see* ρ

Rhodopseudomonas phages: R Φ 1, RS 1

RI (*Nocardia restrictus*): 75 nm diameter head and 10 × 330 nm flexuous tail	Syphoviridae

RS 1 (*Rhodopseudomonas spheroides*): 65 nm diameter head, 60 nm tail, with end plate and fibers, buoyant density in CsCl 1.50 g/cm^3, DNA of 33 × 10^6 dalton

RZh (*Streptococcus*): 90 nm diameter head, 205 nm tail	Myoviridae
R Φ-1 (*Rhodopseudomonas palustris*): lysogenic, 60 nm diameter head and 270 nm tail	Syphoviridae

S 1 (*Bacillus*): temperate phage

Phage and host	Family
S 1 (*Synechococcus*): 50 nm head, rigid 140 nm tail (353 S, buoyant density in CsCl 1.50 g/cm^3), 24 × 10^6 dalton DNA, 13 structural proteins (cyanostylovirus)	Cyanophage
S 2 (*Hemophilus influenzae*): (related to HP 1), particles of 49 × 46 nm with 19 × 117 nm tail, contains 10 × 10^6 dalton DNA with cohesive ends	Myoviridae
S 2 (*Clostridium sporogenes*): similar to F1	Syphoviridae
S-2L (*Synechococcus*)[47]: 56 nm diameter head and flexible 120 nm tail with short thin terminal fiber, contains 2,6 diaminopurine instead of adenine (cyanostylovirus)	Cyanophage
S 13 (*Escherichia coli*)[26]: serological related to ΦX 174	Microviridae
S 708 (*Brucella* spp.): similar to Tbilisi phage	
SA-1 (*Escherichia coli*)	Microviridae
Salmonella Newport: six phages, three bacilliform (dimensions 2.4/1) with contractile tail, two with very long heads (3.5/1) and short tail, one with dubious tail	
SBX-1 (*Xanthomonas*, growing in soybeans): the polyhedral head is elongated (80 × 83 nm) and carries a tail of 112 nm, with base plate and spikes	
SD (*Escherichia coli*): (subgroup II)	Leviviridae
sd (*Escherichia coli* SK): icosahedral head of 59 nm diameter and short (20 × 20 nm) tail (780 S, buoyant density in CsCl 1.45 g/ml) and 43% DNA of 58 × 10^6 daltons (51 S)	Podoviridae
SD 1 (*Pseudomonas*): 50 nm head and 188 nm tail, buoyant density in CsCl 1.52, DNA of 66 × 10^6 daltons	Syphoviridae
Serratia phages: Mp, SM 2, SM 4, SMP, t, η, κ, μ	
SF$_5$ (*Bacillus subtilis*): morphologically similar to Φ 29	
SH 3, SH 5, SH 10, SH 13 (*Streptomyces* phages with DNA with cohesive ends)	
SH 6 (*Streptomyces*): similar to SH 133	Podoviridae
SH 133 (*Hydrogenomonas facilis*): 58 nm head, 29 nm tail	Podoviridae

Phage and host	Family

Shigella phages: Dd, N 17

Si 1 (*Spirillum itersonii*): large icosahedral phage with 26×10^6 dalton DNA

SM-1 (*Synechococcus, Microcystis*): 67 nm head, very short collar with thin appendages (820 S, buoyant density in CsCl 1.48 g/cm³), DNA of 56 $\times 10^6$ daltons, 12 structural proteins (cyanopodovirus) — Cyanophages

SM-2 (*Synechococcus, Microcystis*): 53 nm head, 135 nm tail with curly fibers (cyanostylovirus) — Cyanophages

SM 4 (*Serratia narcescens*): similar to P 22, with complex tail — Myoviridae

SM(B)2 (*Serratia narcescens*): possibly identical with SMP

SMP (*Serratia narcescens*, as well as *Salmonella, Escherichia*): Very large phages, with a head of 135 nm diameter (almost twice the volume of T 2) and a contractile tail of 28 × 235 nm. The tail in extended form resembles the stacked disk form of TMV protein. — Myoviridae

SMP 2 (Enterobacteria): (T-even group) — Myoviridae

SP (*Ancalomicrobium*): 58 nm diameter head and about 120 nm tail lacking plate and fibers — Syphoviridae

SP (*Escherichia coli*): (subgroup IV) — Leviviridae

SP 3 (*Bacillus subtilis*): 115 nm diameter head, 22 × 290 nm tail contracting to 150 nm, DNA of 150 $\times 10^6$ daltons, 5-hydroxymethyluracil instead of thymine, type species — Myoviridae

SP 5 (C), SP 6, SP 7, SP 9 (*Bacillus* spp.)[46]: all contain hydroxymethyluracil instead of thymine

SP 8 (*Bacillus subtilis*)[4,6]: 94 nm diameter head with 152 nm tail, contracting to 130 nm, 69 × 10^6 dalton DNA of 109×10^6 daltons containing hydroxymethyluracil instead of thymine, type species — Myoviridae

SP 10 (*Bacillus subtilis*): pseudotemperate, 90 nm diameter head with 165 nm tail

Phage and host	Family

SP 15 (*Bacillus subtilis*)[4,6]: 120 nm diameter head
with 250 nm tail; many peculiarities, such as
DNA (250 × 10⁶ daltons) with A/G/T/C =
29/21/17/21, with 12% 5-(4′,5′-dihydroxypentyl)-
uracil. The DNA also contains alkali-sensitive
ester bonds and glucose. Density is unusually
high (1.761 g/ml) and T_m low (61.5°). The phage
infects only flagellated strains of bacteria (SP8
group).   Myoviridae

SP 24 (*Streptococcus*): DNA of 42 × 10³ bases, terminally redundant and
circularly permuted

SP 50 (*Bacillus subtilis*)[4]: 88 nm diameter head
with visible capsomers and 25 × 203 nm complex
tail, 108 × 10⁶ dalton DNA (54 S) with breaks in
both strands (related to Φ 1, Φ 2, Φ 14), type species   Myoviridae

SP 60 (*Bacillus subtilis*): 130 × 10⁶ dalton DNA (63 S) with hydroxymethyluracil instead of thymine

SP 70, SP 80 (*Bacillus subtilis*): related to PBS 1

SP 82 (G) (*Bacillus subtilis*)[48]: 100 nm diameter
head and 20 × 165 nm tail, DNA of 130 × 10⁶
daltons containing hydroxymethyluracil instead
of thymine, related to Φ 1, 2C   Myoviridae

SP 90, SP 100 (*Bacillus subtilis*): related to PBS 1

SP 105 (*Bacillus subtilis*): 63 S DNA

SP 272 (Streptococcus): Related to SP 24

SPβ (Bacillus subtilis)[49]: unclassified

Spirillium phages: Si 1

Spiroplasma phages: SV-C1, SV-C2, SV-C3, spv-1, spv-2

SPO 1 (*Bacillus subtilis*)[4,6,37,48]: 100 nm diameter
head, 200 nm tail (base plate, tail tube, 140 nm
sheath), DNA of 87 × 10⁶ dalton (54 S), contains
hydroxymethyluracil instead of thymine (SP 8
group)   Myoviridae

SPO 2 (*Bacillus subtilis*)[6]: temperate phage related
to 105, 50 nm diameter head, 177 nm tail with
complex six-pronged tail tip structure, DNA of 25
× 10⁶ daltons with cohesive ends (φ 105 group)   Syphoviridae

Phage and host	Family
SPP 1 (*Bacillus subtilis*): Closely related to SPO 2. Hexagonal head of 59 nm diameter with 156 nm long tail. DNA of 27×10^6 daltons with cohesive ends. Its strands can be separated on the basis of different densities (1.713 and 1.725 g/ml) in CsCl, and contain, respectively 43 and 57% pyrimidines. This is the most infective DNA known, $5-6 \times 10^3$ molecules being able to initiate a plaque, type species.	Syphoviridae
SPR (*Bacillus subtilis*): related to SP β, SPR, Z, ρ 11, Φ 3 T	
spv-1, spv-2 (*Spiroplasma* spp.): similar or identical with SV-C3	
SPX (*Bacillus subtilis*): defective phage	
SP α (*Bacillus subtilis*): defective phage	
SP β (*Bacillus subtilis*): temperate phage with 85 nm head and 339 nm tail, DNA of 76×10^6 daltons, type species (members: Z, SPR, Φ 3 T, δ 11)	Syphoviridae
ST (*Escherichia coli*): (subgroup III)	Leviviridae
st 1 (*Rhizobium trifolii*)	
St-1 (Enterobacteria)	Microviridae
ST 2	Microviridae
Staphylococcus phages: K, P 11, P 11-M 15, P 52A, SA, Twort, 3A, 3B, 3C, 5C, 6, 11, 44A, 52, 55, 70, 77, 80, 81, 91, 107, 187, 230, 581, 594n (and many more unlisted)	
Streptococcus phages: A 25, AT 298, C 31, Dp 4, GT 234, K 1, M 1, MSP 8, P-a-1, PK 1 R 1, R 2, SH 3, SH 5, SH 8, SH 13, RZh, VD 13, VP 11, 3 ML, 24, 119, 182, 227, Φ 42, Φ 227 (and others not listed)	
Streptomyces phages: VP 5, R 2, MSP 8, Φ 17	
SV-C1 (*Spiroplasma* spp.): 12×250 nm (possibly plectrovirus)	Inoviridae
SV-C2 (*Spiroplasma* spp.)	Possibly Syphoviridae
SV-C3 (*Spiroplasma* spp.): buoyant density in metrizamide 1.26, in CsCl 1.45 g/cm^3, DNA of 14×10^6 daltons, five proteins (similar to MVL 3)	Probably Podoviridae
SW (*Bacillus subtilis*)[6]: head of 101 nm diameter with long contractile tail (22×172 nm, 910 S,	Myoviridae

Phage and host	Family

SW (*Bacillus subtilis*)[6]: (*continued*)
buoyant density in CsCl 1.52 g/cm³), linear DNA
of 62 S, 127 × 10⁶ daltons, the thymine is replaced
by 5-hydroxymethyluracil (SP 8 group)

SW (*Eschericahi coli*): (subgroup II)   Leviviridae

SYPHOVIRIDAE (formerly termed styloviridae, and corresponding to Bradley's group B): Phages of elongated or isometric heads and variously
long noncontractile tails. The only now classified genus is the λ phage
group. Besides λ, the type species and prototype of lysogenic phages,
there is PA 2, Φ D 328, and Φ 80. Strains of λ are Φ 82, Φ 432, Φ 21,
and Φ 424 in decreasingly close relationship.

    Other Syphoviridae are not lysogenic, e.g., T 1. Many others, not
yet classified in terms of genera are, with isometric heads, N4, P22, sd,
Ω8, 7480b (enterobacteria), PIIBNV6-C (*Agrobacterium*), Tb (*Brucella*),
HM2 (*Clostridium*), Cl (*Micrococcus*), gh-1 (*Pseudomonas*), Φ2042, 2
(*Rhizobium*), Φ17 (*Streptomyces*), 114 (*Thermomonospora*), 7-11 (en-
terobacteria), GA-1, Φ29 (*Bacillus*), 182 (*Streptococcus*)

T (*Bacillus megaterium* 899a)

t (*Serratia narcescens*): temperate phage

T 1 (*Escherichia coli*): head of 50 nm diameter, with   Syphoviridae
a long noncontractile tail (10 × 150 nm), the DNA
(34 S) of about 30 × 10⁶ daltons is terminally redundant, without circular permutation, and lacking cohesive ends

T 2 (*Escherichia coli*)[4,37,48] : the hydroxymethyl-   Myoviridae
cytosine is 69% α-glucosylated, 6% β glucosyl-α-
glucosylated (*see* T-even phage group)

T 3 (*Escherichia coli*)[4,48]: closely related and very   Podoviridae
similar to T 7

T 4 (*Escherichia coli*)[4,37,48]: the hydroxymethyl-   Myoviridae
cytosine is 70% α- and 30% β-glucosylated (*see*
T-even phage group) (Figure 34)

T 5 (*Escherichia coli*)[4,37,48]: The head is 65 nm in   Syphoviridae
diameter. The tail (10× 180 nm) consists of about
45 turns of 3 nm pitch. Tail fibers and base plate
are very fine and thus poorly characterized. The
DNA (49 S) is 75 × 10⁶ daltons (69% of the particle weight). It contains no single-strand nicks
but alkali-labile short ribonucleotide sequence.
There are at least 13 proteins, five making up the

Phage and host	Family

T 5 (*Escherichia coli*)[4,37,48]:   (*continued*)
head. The main capsid protein (65%) has a molecular weight of 32×10^3 and minor components
of 43, 30, 28, 23, 19, and 18×10^3 daltons. The
major tail protein (17%) is of 51×10^3, the minor
components of 140, 128, 125, 82, and 70×10^3
daltons. Only 8% of the phage's DNA is initially
injected into the host, the rest only after comple-
tion of the early proteins 2–3 min later (Figure
35).

T 6 (*Escherichia coli*)[4,48]: the hydroxymethylcy-   Myoviridae
tosine is 3% α-glucosylated, 72% β-glucosylated
(*see* T-even phage group)

T 7 (*Escherichia coli*)[4,26d,37,48,50]: Type species of   Podoviridae
the T 7 genus of virulent phages which includes
the very similar T 3. The head is 65 nm in diameter, the 17 nm tail has six short fibers (507 S,
buoyant density in CsCl 1.50 g/cm³). The DNA
is 24×10^6 daltons; it is terminally redundant,
but not circularly permuted. There are about 13
proteins (Figure 36).

Ta (*Thermoactinomyces vulgaris*): 56 nm diameter   Podoviridae
head (519 S), DNA of 29×10^6 daltons

Tbilisi (Tb) (*Brucella abortus*): 65 nm diameter head   Podoviridae
(related to A 422, M 51, S 708, and many more)

TECTIVIRIDAE[51]: Icosahedral 65 nm virions, at times with long spikes at
the vertices (390 S, density in CsCl 1.28 g/cm³) with an internal lipid-
rich mantle (10% of particle weight phospholipid, 5% neutral lipids).
The double-stranded DNA of 9×10^6 daltons is linear (15%). There
are about 20 proteins. The broad-range plasmid-dependent phages adsorb by means of a tail-like structure to the tips of pili of enterobacteria,
Acinetobacter, *Pseudomonas*, *Vibrio Bacillus*, etc. The type species is
PRD 1 and members L 17, PR 3, PR 4, PR 5, PR 772, AP 50, B am 35,
and Φ NS 11.

T-EVEN PHAGE GROUP (GENUS) OF MYOVIRIDAE: (Figure 34) Virions with icosahedral but elongated heads (80×95 nm) and tails of about 16×110
nm with a neck, collar, contractile sheath over the central tube, and a
complex base plate with six spikes and six long fibers (1040 S, buoyant
density in CsCl 1.49 g/cm³). The linear DNA of 130×10^6 daltons (60
S, 54-nm long) is terminally redundant and circularly permuted. It contains hydroxymethylcystosine instead of cytosine, largely α-glucosy-

Phage and host   Family

T-EVEN PHAGE GROUP (GENUS) OF MYOVIRIDAE: (*continued*)

lated and partly carrying glucosyl-glucose. There are about 30 capsid plus tail proteins, as well as two or three internal proteins, and a total of well over 100 genes, many of which are for enzymes and/or for regulatory proteins. Infection of enterobacteria is by attachment of the tail fibers and piercing of the cell wall by the spikes as the sheath contracts. The host's nucleic acid is cannibalized for the synthesis of concatemeric phage progeny DNA. Heads, fibers, and tails are separately assembled. The cell is lysed within 20–30 min.

Some of the phages' lysozymes (as well as other proteins) have been sequenced. T 2 and T 4 lysozyme (gene e) differ in only three amino acids. It functions in lysis from within, while another lysozyme associated with the base plate acts upon infection and lysis from without. The role of the presence of dihydrofolate reductase and pteroylhexaglutamate in the base plate is not understood, nor that of about 140 firmly bound Ca^{2+} ions and ATP molecule in the contractile sheath (which consists of 144 protein molecules), the latter being released upon contraction. There are very many members of this group besides T2, T 4, and T 6, mostly not listed.

Tg (9, 10, 13) (*Bacillus thuringiensis*)

TH 1 (Enterobacteria): (subgroup II)   Leviviridae

Thermoactinomyces phages: Ta, M_1

Thermomonospora phages: 108/106, 114, 119

Thermus thermophylus phage: ΦYS 40

TP 1 C (*Bacillus stearothermophilus*): temperate phage

TP 8, TP 12 (*Bacillus stearothermophilus*): temperate phage

TP 50 (*Bacillus*)[8]: (SP 50 group)   Myoviridae

TP 84 (*Bacillus stearothermophilus*): 37 × 64 nm head, 10 × 150 nm tail (436 S), DNA (30 S) of 32 × 10^6 daltons

TSP-1 (*Bacillus subtilis*)[8]: 90 nm diameter head   Myoviridae
    and 200 nm tail, DNA of 56 × 10^6 daltons, temperature requirement >50° (SP 50 group)

TW 18 (*Escherichia coli*): (subgroup III)   Leviviridae

TW 19, TW 28 (Enterobacteria): (subgroup IV)   Leviviridae

Twort (*Staphylococcus*): 91 nm diameter head, 203   Myoviridae
    nm tail, type species, many members

Type F (*Bacillus subtilis*): 50 × 76 nm head and 158   Syphoviridae
    nm tail

Phage and host	Family
T Φ 3 (*Bacillus stearothermophilus*): DNA (38 S) of 23×10^6 daltons has different buoyant density in CsCl of the two strands	
U 3 (*Escherichia coli*, strain K 12): 22 nm diameter (83 S), thus smaller than Φ X 174, requires cell wall galactose	possibly Microviridae
UC-1 (*Escherichia coli*): 45 nm diameter head and 10 × 150 nm very flexible tail with no terminal features	Syphoviridae
UX (*Bacillus subtilis*): unclassified	
v 1, v 2, v 4, v 5, v 7 (*Vibrio*)	Plasmaviridae
V 6 (*Vibrio parahaemolyticus*): (inovirus)	Inoviridae
V 12, V 14 (*Vibrio parahaemolyticus*): unclassified	
V 45 (*Vibrio foetidus*): 50 nm diameter head, 7 × 24 nm tail	
VA (*Ancalomicrobium*): 38 nm diameter head and 104 nm tail lacking terminal features	Syphoviridae
VA-1 (*Vibrio cholerae* NIH 41)	Myoviridae
VD 13 (*Streptococcus*): 43 × 113 nm head, 145 nm tail	Syphoviridae
Vi I (Enterobacteria): isometric, with fibers at the collar	Myoviridae
Vi II (Enterobacteria): isometric	Syphoviridae
Vibrio phages: v 6, v 12, v 14, v 45, VA 1, O6N58P, 149, Φ 2	
VK (*Escherichia coli*): (subgroup III, very similar to Qβ)	Leviviridae
VL-1 (*Bdellovibrio*): about 50 nm diameter head and 50 nm tail	possibly Myoviridae
VP 1 (VP 7, VP 18), (*Vibrio parahaemolyticus*)[6]: 90 nm diameter head, 97 nm tail (neck appendages)	Myoviridae
VP 3 (VP 6), (*Vibrio parahaemolyticus*): 69 nm diameter head, with knoblike projections, 233 nm tail, (Figure 37)	Syphoviridae
VP 5 (*Streptomyces*): 59 nm diameter head, 166 nm tail	Syphoviridae
VP 5 (VP 15, VP 16), (*Vibrio parahaemolyticus*)[6]: 52 × 92 nm diameter head, 156 nm tail	Syphoviridae

Phage and host	Family
VP 11 (*Streptomyces coelicolor*): 55 nm diameter head, 11 × 220 nm tail	Myoviridae
VP 11 (*Vibrio parahaemolyticus*)[6]: 55 × 72 nm head, 138 nm tail, 55 × 10⁶ dalton DNA	Syphoviridae
VP 12 (VP 13), (*Vibrio parahaemolyticus*)[6]: 78 nm diameter head, 167 nm tail	Syphoviridae
Vx (*Bacillus*)[8]: (SP 8 group)	Myoviridae
W (*Micrococcus luteus*)[10]: related to 186	Syphoviridae
W (*Excherichia coli*): related to T 3, T 7	Podoviridae
W 3, W 8 (*Streptococcus pneumoniae*)[10]: 60 nm head, 10 × 180 nm tail, DNA of 33 × 10⁶ daltons	Syphoviridae
W 14 (*Pseudomonas acidovorans*): 85 nm head, 140 nm tail, DNA of buoyant density in CsCl 1.666 g/cm³, T_m 99.3°C (73% G + C). About half of thymine replaced by 5-(4-aminobutylaminomethyl)-uracil i.e., thyminyl-putrescine.	Myoviridae
W 31 (*Escherichia coli*): similar to Φ I, Φ II	Podoviridae
WA/1 (*Escherichia coli*)	Microviridae
WAK/2 (*Escherichia coli*): long tail with forked tip	Syphoviridae
Wb (*Brucella*)	Podoviridae
WF/1 (*Escherichia coli*)	Microviridae
WLL (*Excherichia coli*): phage used by Schlesinger in classical virus-characterizing studies about 1934	
WT 1 (*Rhizobium trifolii*): 64 nm head, 128 nm tail	Myoviridae
X (*Micrococcus luteus*): related to 186	Syphoviridae
X 1 (*Salmonella typhimurium*): 66 nm diameter head, 14 × 220 nm tail	
X 2–6 (*Salmonella typhimurium*): unclassified	
X 7 (*Salmonella typhimurium*): 85 nm diameter head, 19 × 263 nm tail and 185 nm fibers	
X 29 (*Vibrio*)[6]: 64 nm head, 142 nm tail	Myoviridae
X 174: *see* Φ X 174	
Xanthomonas phages: Cf, CBX-1, Xf, XP 5, XP 12	
Xf, XF 2 (*Xanthomonas oryzae*): coat protein with 44 aminoacids (inovirus)	Inoviridae

Phage and host	Family
XP 5 (*Xanthomonas prunii*): 60 nm diameter head, 88 nm tail	Myoviridae
XP 12 (*Xanthomonas*): 55 × 78 nm head and 142 nm tail, DNA of 34 × 10^6 daltons that contains 5-methylcytosine instead of cytosine	Syphoviridae
Y (Enterobacteria)	Podoviridae
Z (*Bacillus subtilis*): temperate phage related to SPR, SPβ, Φ 3 T, and ρ 11	
ZG/1 (*Escherichia coli*): (subgroup IV)	Leviviridae
ZG/2 (Enterobacteria)[26d]: (inovirus)	Inoviridae
ZG/3A (Enterobacteria): elongated head	Syphoviridae
ZIK/1 (*Escherichia coli*): probably subgroup II, coat protein of 12 × 10^3 daltons, lacks histidine, methionine, cysteine	Leviviridae
ZJ 1 (*Escherichia coli*): (subgroup I)	Leviviridae
ZJ/2 (Enterobacteria): closely related to fd	Inoviridae
ZL/3 (*Escherichia coli*): (subgroup IV)	Leviviridae
ZR (*Escherichia coli*): (subgroup I, closely related to MS 2)	Leviviridae
ZS/3 (*Escherichia coli*): (subgroup IV)	Leviviridae
α (*Bacillus tiberius* and *megaterium*)[8]: temperate isometric phage with 61 nm head and 121 nm tail (470 S) containing 35 × 10^6 dalton DNA (40 S) with an average of one interruption per chain. This was the first DNA virus in which it was found possible to separate the strands by density gradients (1.717 and 1.724 g/cm^3 in CsCl).	Syphoviridae
α1 (Enterobacteria): (T-even group)	Myoviridae
α1, α2 (*Achromobacter*): closely related to α 3, though different tail lengths	Microviridae
α1, α2 (*Clostridium botulinum*): different buoyant density in CsCl for the two strands	
α3 (*Achromobacter*, sp 2)[52]: general transducing phage, 45 nm diameter head, 320 nm tail, buoyant density in CsCl 1.506 g/cm^3, DNA 34 × 10^6 daltons.	Syphoviridae

Phage and host	Family
α3 (*Escherichia coli*)	Microviridae
α3A (*Vibrio*)[6]: 48 × 66 nm head, 330 nm tail, DNA of 34 × 10^6 daltons	Syphoviridae
α10 (Enterobacteria)	Microviridae
α15, α17 (*Escherichia coli*): (probably subgroup IV)	Leviviridae
β (*Corynebacterium diphtheriae*): 55 × 61 nm head and 10 × 287 nm tail (related to ω)	Syphoviridae
β (*Escherichia coli*) (subgroup I)	Leviviridae
β3 (*Bacillus subtilis*): pseudotemperate phage, DNA of 59 × 10^6 daltons containing an unidentified base and terminally-bound protein (related to M 2)	Podoviridae
β4 (Enterobacteria): isometric	Syphoviridae
β22 (*Bacillus subtilis*): large virulent phage unrelated to SP 82, 100 nm diameter head and 220 nm tail	
γ (*Bacillus anthracis*): possibly related to Φ 80	
γ (*Corynebacterium diphtheriae*): closely related to β	
γ(2) (*Escherichia coli*)	Syphoviridae
δ1 (Enterobacteria)	Microviridae
δA (Enterobacteria): (fd group of inoviruses)	Inoviridae
ε15, ε34 (*Salmonella anatum*): temperate phages	
ζ3 (Enterobacteria)	Microviridae
η (*Serratia narcescens*): 73 × 10^6 particle weight, buoyant density in CsCl is 1.495 g/ml, DNA of 100 × 10^6 daltons with part of the guanine being replaced by a not yet identified base of an absorbance maximum at all pHs above 280 nm; an unidentified sugar is also present.	
η8 (Enterobacteria): related to λ, Φ 80	Microviridae
θ1 (*Bacillus licheniformis*): Oblong head of 50 × 100 nm and thin tail of 200 nm with base plate and spikes; related to LP 52 (Figure 38)	Syphoviridae
κ (*Serratia narcescens*): temperate phage, the DNA of 100 × 10^6 daltons shows, in contrast to η, no unusual components	
κ (*Vibrio*): 61 nm head, 111 nm tail	Myoviridae

Phage and host	Family

λ (*Escherichia coli*) (Figure 39) Type species of genus λ phage group, the classical prototype of lysogenic (temperate) phages. Icosahedral particles of 54 nm diameter with evident capsomers, carrying a thin noncontractile tail (15 × 150 nm) with fine cross-striation and a thin fiber (2 × 25 nm) at the end. The particle weight is $60 × 10^6$ (416 S), the buoyant density in CsCl 1.49 g/cm^3. The phage contains one molecule of double-stranded DNA (34 S) of $30 × 10^6$ daltons and 17.2 μm contour length. The composition is 49% (G+C). The DNA has cohesive ends (12 nucleotide pairs long). It yields upon shearing two halves of unequal (G+C) content, the so-called left-hand portion being denser and higher in (G+C) than the right (55 vs. 45%). There is also a difference in the density of the two strands, but this difference is less than in many other phages.

Syphoviridae

The main proteins have been identified as composing the head ($38 × 10^3$ daltons, 60% or 540 molecules), the tail ($31 × 10^3$ daltons, 19%), the tail fiber ($130 × 10^3$ daltons), and a protein playing an internal role, particularly during maturation ($12 × 10^3$, 19%, 550 molecules per particle). Mutant capsids, termed "petit," lack that protein. There are also six minor protein components ranging from 79 to $14 × 10^3$ daltons. At least 18 genes are involved in the morphogenesis of λ, seven with head formation and the rest with tail assembly. The endolysin, the product of gene R, has been purified and sequenced (157 residues, one cysteine). It is unrelated to the T-even phage lysozymes. Most of the DNA has actually been sequenced.

Many more or less closely related strains of λ are known. These are, in order of diminishingly close relationship, 82 and 434, 21, 424, and 80. They are listed separately in this catalogue. (Phage 186 is related to P 2, 299, D, and N 1, not to λ). These and others are defective mutants, carrying varying amounts of the host's *gal* and *bio* gene. Their DNAs differ in molecular weight, contour length, and composition.

μ (*Serratia narcescens*): probably identical to η

Phage and host	Family
μ, μ1 (also termed Mu), (*Escherichia coli, Salmonella*, etc.)[55]: A temperate phage of appearance and size similar to λ but in density and DNA content more similar to defective λ strains; serologically not related to λ. The DNA is 12.9-μm long, of 30.7 S, and 25 × 10^6 daltons. The phage has a unique propensity for stable integration at many positions, and for increased mutation frequency.	Syphoviridae
μ2 (*Escherichia coli*): (subgroup II)	Leviviridae
μ4 (*Bacillus stearothermophilus*): 55 nm diameter head and 7 × 225 nm tail lacking base plate and fibers, DNA of 33 S, 26 × 10^6 daltons, the strands can be separated on CsCl gradients	Syphoviridae
μ4 (*Bacillus subtilis*): unusually small (10 nm) tailless particle	
o6 (Enterobacteria)	Microviridae
ρ11 (*Bacillus subtilis*): 82 nm diameter head, 380 nm tail, 11 nm base plate with pins (temperate phage related to Φ 3 T and SPR)	Syphoviridae
Φ1, Φ2 (*Bacillus subtilis*)[8]: related to SP 50	
Φ2 (*Vibrio cholerae*)[6]: DNA of 113 × 10^6 daltons, five proteins	
Φ3, Φ4 (*Escherichia coli*): female-specific host range mutants of ΦW	
Φ3T (*Bacillus subtilis*)[56]: temperate phage related to SPβ, ρ11, 81 nm diameter head and 11 × 380 nm tail with 32 nm base plate with prongs	Syphoviridae
Φ6 (*Caulobacter*): requires pilus for infections	Syphoviridae
Φ6 (*Pseudomonas phaseolicola*)[57]: The phage shows a polyhedral 60 nm head covered by a lipid envelope, the lipid composition resembling that of the host. The tail is short and complex. Lipid content 25%, RNA 13%, protein 62%. Buoyant density in CsCl 1.27 g/cm^3. Three double-stranded RNAs of molecular weights 1.9, 2.8, and 4.6 × 10^6. The phage is sensitive to ether, etc. It contains RNA polymerase; type species.	Cystoviridae
Φ14 (*Bacillus subtilis*): (SP 50 group)	Myoviridae
Φ15 (*Bacillus subtilis*)[8]: (Φ 29 group)	
Φ17 (*Streptomyces*)[10]: 60 nm diameter head and 14 nm tail	Podoviridae

Phage and host	Family

Φ21: *see* 21

Φ23-1-a (*Pseudomonas phaseolicola*): insensitive to organic solvents

Φ25 (*Bacillus subtilis*)[8]: 75 nm diameter head and   Myoviridae
13 × 130 nm tail, 64 S DNA containing hydroxymethyluracil instead of thymine (SP 8 group)

Φ29 (*Bacillus subtilis*)[4,37,58]: 32 × 46 nm head, 6 × 36 nm tail, 14 nm
fibers on head and collar (256 S). Its 12 × 10⁶ dalton DNA, linear
and not permuted, carries a 5'-terminal protein (31 × 10³ daltons).
Both the head and the neck collar with appendages consist of three
proteins, and the tail of one protein (related to Φ15, NF, GA-1, SF 5),
(Figure 40).

Φ42 (*Streptococcus*, Group H)[59]: temperate phage, DNA of 25 × 10⁶
daltons, circularly permuted and terminally redundant

Φ80 (*Escherichia coli*): deletion mutant of λ with   Syphoviridae
same cohesive ends, 8% less DNA

Φ105 (*Bacillus subtilis*)[8]: temperate phage, 59 nm   Syphoviridae
diameter head, 10 × 200 nm tail, 26 × 10⁶ dalton
DNA, type species, note narrow neck, killer particle (Figure 41)

Φ149 (*Vibrio cholorae*) (566 S) Contains linear DNA   Podoviridae
of 36 × 10⁶ daltons

Φ186: related to P 2

Φ227 (Group H) (*Streptococcus sanguis*)[10]: Tem-   Syphoviridae
perate phage with 56 × 52 nm head, 10 × 149
nm tail, buoyant density in CsCl 1.50 g/cm³,
DNA of 23 × 10⁶ daltons

Φ2037/1-Φ2037/7 (*Rhizobium*)[60]: 77 × 147 nm   Syphoviridae
head, 261 nm tail

Φ2042 (*Rhizobium*)[60]: 64 × 75 nm diameter head,   Podoviridae
17 nm tail

Φ2193/2 (*Rhizobium*)[60]: 108 nm diameter head, 23   Myoviridae
× 147 nm tail

Φ2205; related to Φ2037/1

ΦI (*Escherichia coli*): related to T 7   Podoviridae

ΦII (*Escherichia coli*): female-specific, similar to T   Podoviridae
7

ΦA (*Escherichia coli*): serologically related to ΦX   Microviridae
174

Phage and host	Family
ΦA1 (*Bacterioides fragilis*): 63 nm head and 153 nm tail, lacking a sheath	Syphoviridae
ΦAG 8010 (*Arthrobacter globiformis* 8010)[61]: 60 × 69 nm head, 120 nm sheathless tail	Syphoviridae
ΦC (*Actinomyces*): 52 nm diameter head and 92 nm tail	Syphoviridae
ΦC (*Escherichia coli*)	Microviridae
ΦC (Nocardia)	Syphoviridae
ΦCb5, Cb8, Cb8r, Cb12r, Cb23r (*Caulobacter*)	Leviviridae
ΦCb 13 (*Caulobacter*): bacilliform with short tail	
ΦCbK (*Caulobacter crescentus*)[37,62]: Bacilliform (64 × 195 nm), with flexible tail of 275 nm, tail fibers, and head fibers	Syphoviridae
ΦCL 29 (CLV 29), (*Caryophanum latum*): 70 × 130 nm head, 6.7 × 330 nm tail, buoyant density in CsCl 1.450 mg/cm^3	Syphoviridae
ΦCP (*Actinomyces*): strain of ΦC	
ΦCP 2 (*Caulobacter crescentus*): 29 nm diameter	Leviviridae
ΦCP 18, ΦCP 28, ΦCP 32, ΦCP 42 (*Caulobacter*): serologically related to ΦCP 2	Leviviridae
ΦCR (*Actinomyces*): strain of ΦC	
ΦCR 1–ΦCR 13, ΦCR 15–ΦCR 27, ΦCR 29–ΦCR 38 (*Caulobacter crescentus*): 60–200 nm heads, 50–330 nm tails	
ΦCr 14, ΦCr 28 (*Caulobacter*): 22 nm diameter, requires pili	Leviviridae
ΦD 326 (*Escherchia coli*): λ phage group	Syphoviridae
Φe (*Bacillus subtilis*)[8]: (SP 8 group)	Myoviridae
ΦK: *see* K	
ΦK 2 (*Pseudomonas*)	Myoviridae
ΦKZ: 130 nm diameter head	Podoviridae?
ΦLL 55 (*Lactobacillus lactis*): 50 nm diameter head and 7 × 200 nm tail	Syphoviridae
ΦNR2rH: DNA of 92 × 10^6 daltons	

Phage and host	Family
ΦNS 11 (*Bacillus*): 68 nm diameter head with 9 × 10⁶ dalton DNA	Tectiviridae
ΦPLS-1 (*Pseudomonas aeruginusa*): 70 nm diameter head and 120 nm tail, lipopolysaccharide-specific	Myoviridae
ΦR (Enterobacteria)	Microviridae
ΦRE (Staphylococcus): identical with 6	
ΦT (*Bacillus megaterium* 899a): 68 × 57 nm head, 10 × 240 nm tail with 15 nm terminal disk	Syphoviridae
ΦW (*Escherichia coli*): probably identical with ΦII	
ΦW 14 (*Pseudomonas*): *see* W 14	Myoviridae
ΦW 31 (*Escherichia coli*): related to ΦW	
ΦX 174 (*Escherichia coli*)[26d,63]: type species (Figure 42) (*see* Microviridae)	Microviridae
ΦYS 40 (*Thermus thermophilus*), (extreme thermophile): 125 nm diameter head, 178 nm tail with 27 nm base plate and fibers	probably Myoviridae
Φ_γ (Escherichia coli): related to λ, Φ80	
Φ_μ4: *see* μ4	
χ (*Chondrococcus columnaris*)	Myoviridae
χ (Enterobacteria): Attacks the flagellae of motile Salmonella strains only; 67 nm diameter head, 14 × 220 nm tail with 55 fine striations (pitch 4.2 nm), and a single tail fiber (2.2 × 210 nm).	Syphoviridae

Ψ (*Agrobacterium tumefaciens*, strain B91): temperature-sensitive (<37°), very similar or identical to Ω

ω (*Corynebacterium*): 52 × 57 nm head and 283 nm tail

| ω1, ω2, ω3, ω4, ω8 (*Pneumococcus*): 50 nm diameter head and 200 nm tail with fiber at the end | Syphoviridae |

ω7, ω9 (*Pneumococcus*): different from ω1, etc.

Ω (*Agrobacterium tumefaciens*): (closely related if not identical to LV 1, R 4, PS 8, PB 2A), *see* LV 1

| Ω8 (*Escherichia coli*), (08-specific): 49 nm head and with 14 nm base plate with 4 × 13 nm spikes and 168 nm fibers | Podoviridae |

Phage and host	Family
06 N 58 P (*Pseudomonas*)	possibly Corticoviridae
1 (*Clostridium sporogenes*): similar to F 1	
1 (*Bacillus subtilis*): see Φ 1	
1X1 (*Pseudomonas aeruginosa*): unclassified	
1Φ1, 1Φ3, 1Φ7, 1Φ9 (*Escherichia coli*)	Microviridae
2 (*Salmonella typhimurium*): similar to E 1	
2 (*Rhizobium*)	Podoviridae
2, 6, 7, 8, 9, 10 (*Pseudomonas aeruginosa*): temperate phages	
2C (*Bacillus subtilis*)[8]: 88 nm diameter head, 15 × 142 nm tail. The DNA (100×10^6 dalton) contains hydroxymethyluracil instead of thymine. The strands have large terminal redundancy and very different buoyant density in CsCl (SP82G group).	Myoviridae
2G/2 (*Escherichia coli*): (fd group of inoviruses)	Inoviridae
2G 3A (*Escherichia coli*): elongated head	Syphoviridae
3 (*Bacillus subtilis*): T-even group	Myoviridae
3A (*Staphylococcus*)[10]: 35 × 42 nm head, 307 nm tail, type species (many members)	Syphoviridae
3B (*Staphylococcus* B)[10]: 60 × 80 nm head, 300 nm tail	Syphoviridae
3C (*Staphylococcus*)[10]: bacilliform 45 × 95 nm head, 270 nm thin tail with very thin collar	Syphoviridae
3ML (*Streptococcus*)[10]: 43 × 56 nm head, 96 nm tail	Syphoviridae
3NT (*Bacillus subtilis*): pseudotemperate, related to PBS 1	
3T + (*Escherichia coli*): (T-even group)	Myoviridae
5C (Staphylococcus): probably identical with 2C	
6 (*Pseudomonas phaseolicola*): see Φ 6	
6 (*Staphylococcus* B)[10]: 40 × 92 nm head, and 300 nm tail. (Figure 43)	Syphoviridae
7-7-7 (*Rhizobium*)[64]: very elongated head	Syphoviridae
7-11 (*Salmonella Newport*): 40 × 154 nm head and very short tail[65] (Figure 44)	Podoviridae

Phage and host	Family
7S (*Pseudomonas aeruginosa*, Enterobacteria): related to PP7	
9/0 (*Escherichia coli*): (T-even group)	Myoviridae
9NA (*Salmonella*, smooth strains): 60 nm head, 150 nm tail with 30 nm base plate	Syphoviridae
10/tur (*Escherichia coli*): (plectrovirus)	Inoviridae
11F (Enterobacteria): (T-even group)	Myoviridae
12B (*Pseudomonas syringae*): varying size head and complex tail structure	Myoviridae
12S (*Pseudomonas syringae*): 80 nm diameter head and 100 nm tail	Myoviridae
13 M (*Proteus mirabilis*): similar to 5006	
14 (*Bacillus subtilis*): unclassified	
15 (*Bacillus subtilis*): related to Φ 29	
15 (*Escherichia coli*): 60 nm diameter elongated head, 12 nm neck, 20 × 100 nm tail	Myoviridae
16-6-12 (*Rhizobium*)[14]: isometric 54 nm head, 144 nm tail	Syphoviridae
16-12-1 (*Rhizobium*)[14]: 59 nm head and 12 × 141 nm forked tail, attaches to pili	Syphoviridae
16-19 (*Salmonella Newport*): 48 × 167 nm head, 185 nm tail	Myoviridae
17 (*Actinomyces*): *see* Φ 17	
21 (*Escherichia coli*): related to P 22 and λ with same cohesive ends	Syphoviridae
24 (*Streptococcus*)[10]: 55 nm diameter head, 255 nm tail	Syphoviridae
25: *see* Φ 25	
29: *see* Φ 29	
29α (*Escherichia coli*): related to T 5	
34 (*Proteus mirabilis*): similar to 5006	
41C (*Bacillus subtilis*): 50 nm diameter head and 10 × 140 nm tail lacking base plate and fibers	Syphoviridae
42: *see* Φ 42	
44A (*Staphylococcus aureus* and *pyogenes*): unclassified	

Phage and host	Family
47 (*Staphylococcus*): identical with b	
50 (Enterobacteria): (T-even group)	Myoviridae
52 (*Staphylococcus* B)[10]: elongated 50 nm diameter head with 150 nm tail (52HJD: strain of 52)	Syphoviridae
55 (*Staphylococcus*)	Syphoviridae
66F (Enterobacteria): (T-even group)	Myoviridae
66t (*Salmonella typhimurium*): (T-even group)	Myoviridae
70 (*Staphylococcus* B)[10]: 53 × 98 nm head and 300 nm tail	Syphoviridae
75 (*Bacillus pumilis*): unclassified	
77 (*Staphylococcus* B)[10]: 55 nm head and 220 nm tail, type species (many members)	Syphoviridae
80 (*Clostridium perfringens*): (HM2 group)	Podoviridae
80 (*Staphylococcus pyogenes*): unclassified	
81 (*Staphylococcus pyogenes*): unclassified	
82 (*Escherichia coli*)[66]: similar and related to λ with same cohesive ends	
91 (*Staphylococcus* B)[10]: 50 nm diameter head and 150 nm tail	
105 (*Bacillus subtilis*)[8]: Temperate phage, similar to SPO 2 (slight serological cross-reaction). The head is hexagonal with 52 nm diameter, the tail 10 × 220 nm, equipped with hexagonal end plate, 23 nm in diameter, but no tail fibers. The DNA is of 25 × 10^6 mol. wt., possibly with cohesive ends.	Syphoviridae
107 (*Staphylococcus*)[10]: 58 nm diameter head and 193 nm tail, type species	Syphoviridae
108/106 (*Thermomonospora*): 84 × 110 nm head and 140 nm tail	Myoviridae
114 (*Thermomonospora*): 56 nm diameter head and 17 nm tail	Podoviridae
119 (*Thermomonospora*): 60 nm diameter head and 183 nm tail	Syphoviridae
121 (Enterobacteria): isometric	Myoviridae
143/tur (*Acholeplasma*): (plectrovirus)	Inoviridae

Phage and host	Family
149 (*Vibrio cholerae*): 75 × 85 nm head, flexuous tail of 10 × 220 nm with terminal knob of 13 nm diameter, equipped with hexagonal end plate 23 nm in diameter, but no tail fibers. The DNA is of 25×10^6 mol. wt., possibly with cohesive ends.	Syphoviridae
179/tur (*Acholeplasma*): (plectrovirus)	Inoviridae
182 (*Streptococcus*)[10]: 32 × 40 nm head, 32 nm tail with appendages on neck, type species (many members)	Podoviridae
182/tur (*Acholeplasma*): (plectrovirus)	Inoviridae
182a (*Streptococcus*)	Syphoviridae
186 (*Escherichia coli*)[26d]: temperate phage, related to P 2, P 4, 299, D, and N 1, similar to but not closely related to λ, with different cohesive sequence	Syphoviridae
187 (*Staphylococcus*): 59 nm diameter head, 173 nm tail	Syphoviridae
227: *see* Φ 227	
230 (*Staphylococcus*)	Syphoviridae
299 (*Escherichia coli*): unstable phage, 60 nm diameter head and 140 nm tail, DNA of 21×10^6 daltons with cohesive ends similar to those of P 2	Syphoviridae
317 (*Rhizobium leguminosarum*): 59 nm diameter head and tail with three fibers of 27 nm, 41×10^6 dalton DNA	Syphoviridae
363 (*Escherichia coli*): temperate deletion mutant of λ	
424 (*Escherichia coli*): temperate deletion mutant of λ with same cohesive ends	
434 (*Escherichia coli*): temperate deletion mutant of λ	
525 (*Pseudomonas phaseolicola*): phage with 56 nm diameter head, 15 nm tail with base plate, DNA of 25×10^6 daltons	Podoviridae
581 (*Staphylococcus aureus*): phage with 55 nm head and 240 nm tail	
594n (*Staphylococcus*)[10]: phage with 55 × 96 nm head and 300 nm tail	Syphoviridae

Phage and host	Family
1304c1r (*Acholeplasma*): (plectrovirus)	Inoviridae
1307 (*Mycoplasma*)	Plasmaviridae
1412 (Enterobacteria)	Podovirdiae
3610 (*Bacillus subtilis*): defective phage of various sizes similar to PBSX, α and μ	
4996 (*Vibrio*): 65 nm diameter head, 18 nm tail	Podoviridae
5006(M), (*Proteus mirabilis*)[67]: General transducing phage, 46 nm diameter head, buoyant density in CsCl 1.491, 16 × 16 nm tail, DNA of 21×10^6 daltons circularly permuted and terminally redundant	Podoviridae
5845 (*Escherichia coli*): (T-even group)	Myoviridae
7480b (Enterobacteria)	Podoviridae
8762, 8764 (*Alcaligenes faecalis*): 60 nm diameter head and 170 nm tail	Syphoviridae
8893 (*Alcaligenes faecalis*): (T-even group)	Myoviridae
9266 (Enterobacteria): elongated head	Myoviridae
I (*Vibrio*): 71 nm diameter head, 13 nm tail	Podoviridae
II (*Bacillus*): 97 nm diameter head and 539 nm tail, type species	Syphoviridae
II (*Vibrio*): 64 nm diameter head, 79 nm tail, DNA of 93×10^6 daltons	Myoviridae
III (*Vibrio*): 61 nm diameter head, 16 nm tail	Podoviridae
IV (*Vibrio*): 83 nm diameter head, 159 nm tail, DNA of 68×10^6 daltons	Syphoviridae

References (Section III)

1. Berthiaume, L., and Ackermann, H.-W., 1977, La classification des actinophages, *Pathol. Biol.* **25:**195–201.
2. Ackermann, H.-W., Simon, F., and Verger, J-M., 1981, A survey of *Brucella* phages and morphology of new isolates, *Intervirology* **16:**1–7.
3. Nagy, E., Prágai, B., and Ivánovics, G., 1976, Characteristics of phage AP50, an RNA phage containing phospholipids, *J. Gen. Virol.* **32:**129–132.
4. Matthews, C. K., 1979, Reproduction of large virulent bacteriophages, in: *Comprehensive Virology*, Vol. 7 (H. Fraenkel-Conrat and R. R. Wagner, eds.), p. 179, Plenum Press, New York.
5. Morgan, T. M., and Stanisich, V. A., 1976, Characterization and properties of Phage B33, a female-specific phage of *Pseudomonas aeruginosa, J. Gen. Virol.* **30:**73–79.
6. Ackermann, H.-W. *et al.*, 1984, Classification of *Vibrio* bacteriophage, *Intervirology*, in press.
7. Shaw, A. R., Lang, D., and McCorquodale, D. J., 1979, Terminally redundant deletion mutants of bacteriophage, *J. Virol.* **29:**220–231.
8. Reanney, D. C., and Ackermann, H.-W., 1981, An updated survey of *Bacillus* Phages, *Intervirology* **15:**190–197.
9. Brownell, G. H., and Adams, J. N., 1967, Growth and characterization of nocardiophages for *Nocardia canicruria* and *Nocardia erythropolis* mating types, *J. Gen. Microbiol.* **47:**247–256.
10. Ackermann, H.-W., 1975, La classification des bacteriophages des cocci gram-positifs: *Micrococcus, Staphylococcus* et *Streptococcus, Pathol. Biol.* **23:**247–253.
11. Sobieski, R. J., and Olsen, R. H., 1973, Cold-sensitive *Pseudomonas* RNA Polymerase. I. Characterization of the host-dependent cold-sensitive restriction of phage CB3, *J. Virol.* **12:**1375–1383.
12. Dai, H., Chiang, K.-S., and Kuo, T.-T., 1980, Characterization of a new filamentous phage Cf from *Xanthomonas citri, J. Gen. Virol.* **46:**277–289.
13. Matthews, R. E. F., 1982, Corticoviridae, *Intervirology* **17:**67.
14. Ackermann, H.-W., 1978, La classification des phages d'*Agrobacterium* et *Rhizobium, Pathol. Biol.* **26:**507–512.
15a. Sherman, L. A., and Bron, R. M., Jr., 1978, Cyanophages and viruses of eukaryotic algae, in: *Comprehensive Virology*, Vol. 12 (H. Fraenkel-Conrat and R. R. Wagner, eds.), p. 145, Plenum Press, New York.
15b. Saffermann, R. S. *et al.*, 1982, Classification of viruses of cyanobacteria, *Intervirology* **19:**61.
16. Matthews, R. E. F., 1982, Cystoviridae, *Intervirology* **17:**80.
17. Gill, G. S., Hull, R. C., and Curtiss, R., III., 1981, Mutator bacteriophage D108 and its DNA: An electron microscopic characterization, *J. Virol.* **37:**420–430.
18. Nikolsaya, I. I., Trushinskaya, G. N., and Tikchonenko, T. I., 1972, Some properties of mononucleotides from the DD VII phage DNA, *Dokl. Akad. Nauk. USSR* **205:**241–243.

19. Garcia, E., Ronda, C., and López, R., 1980, Replication of bacteriophage Dp-4 DNA in *Streptococcus pneumoniae*, *Virology* **105**:405–414.

20. Moses, P. B., Boeke, J. D., Horiuchi, K., and Zinder, N. D., 1980, restructuring the bacteriophage f1 genome: Expression of gene VIII in the intergenic space, *Virology* **104**:267–278.

21. Yonesaki, T., Furuse, K., Haruna, I., and Watanabe, I., 1982, Relationships among four groups of RNA coliphages based on the template specificity of GA replicase, *Virology* **116**:379–381.

22. Beck, E., Sommer, R., Auerswald, E. A., Kurz, C., Zink, B., Osterburg, G., Schaller, H., Sugimoto, K., Sugisaki, H., Okamoto, T., and Takanami, M., 1978, Nucleotide sequence of bacteriophage fd DNA, *Nucl. Acids Res.* **5**:4495.

23a. Godson, G. N., Barrell, B. G., Staden, R., and Fiddes, J. C., 1978, Nucleotide sequence of bacteriophage G4 DNA, *Nature* **276**:236.

23b. Sanger, F., Air, G. M., Barrell, B. G., Brown, N. L., Coulson, A. R., Fiddes, J. C., Hutchison, C. A., III, Slocombe, P. M., and Smith, M., 1977, Nucleotide sequence of bacteriophage ϕX174 DNA, *Nature* **265**:687.

23c. Sanger, F., Coulson, A. R., Friedmann, T., Air, G. M., Barrell, B. G., Brown, N. L., Fiddes, J. C., Hutchison, C. A., III, Slocombe, P. M., and Smith, M., 1978, The nucleotide sequence of bacteriophage ϕ174, *J. Mol. Biol.* **125**:225.

24. Auling, G., Bernard, U., Huttermann, A., and Mayer, F., 1980, Characterization and comparison of the DNAs of the three closely related bacteriophages gd, ge, and gf with the genome DNA of the hydrogen-oxidizing host strain *Pseudomonas pseudoflava* GA3, *J. Gen. Virol.* **49**:51–59.

25. Dhillon, T. S., 1981, Temperate coliphage HK022: Virions, DNA, one-step growth, attachment site and the prophage genetic map, *J. Gen. Virol.* **55**:487–492.

26a. Horiuchi, K., Vovis, G. F., and Model, P., 1978, The filamentous phage genome: genes physical structure, and protein products, in: *The Single-Stranded DNA Phages* (D. T. Denhardt, D. Dressler, and D. S. Ray, eds.), pp. 113–137, Cold Spring Harbor Laboratory, Cold Spring Harbor, New York.

26b. Matthews, R. E. F., Inoviridae, *Intervirology* **17**:78.

26c. Marvin, D. A., and Wachtel, E. J., 1975, Structure and assembly of filamentous bacterial viruses, *Nature* **253**:19.

26d. Air, G. M., 1979, DNA sequencing of viral genomes, in: *Comprehensive Virology*, Vol. 13 (H. Fraenkel-Conrat and R. R. Wagner, eds.), p. 205, Plenum Press, New York.

26e. Ray, D. S., 1977, Replication of filamentous bacteriophages, in *Comprehensive Virology*, Vol. 7 (H. Fraenkel-Conrat and R. R. Wagner, eds.), p. 105, Plenum Press, New York.

27. Sozzi, T., Watanabe, K., Stetter, K., and Smiley, M., 1981, Bacteriophages of the genns *Lactobacillus*, *Intervirology* **16**:129–135.

28. Rees, P. J., and Fry, B. A., 1981, The morphology of staphylococcal bacteriophage K and DNA metabolism in infected *Staphylococcus aureus*, *J. Gen. Virol.* **53**:293–307.

29. Wollin, R., Eriksson, U., and Lindberg, A. A., 1981, *Salmonella* bacteriophage glycanases: Endorhamnosidase activity of bacteriophages P27, 9NA, and KB1, *J. Virol.* **38**:1025–1033.

30a. Matthews, R. E. F., 1982, Leviviridae, *Intervirology* **17**:136.

30b. Fiers, W., Contreras, R., Duerinck, F., Haegeman, G., Iserentant, D., Merregaert, J., Jou, W. M., Molemans, F., Raeymaekers, A., Van den Berghe, A., Volckaert, G., and Ysebaert, M., 1976, Complete nucleotide sequence of bacteriophage MS2-RNA: Primary and secondary structure of the replicase gene, *Nature* **260**:500.

31. Joshi, A., Siddiqi, J. Z., Rao, G. R. K., and Chakravorty, M., 1982, MB78, a virulent bacteriophage of *Salmonella typhimurium*, *J. Virol.* **41**:1038–1043.

32a. Matthews, R. E. F., 1982, Microviridae, *Intervirology* **17**:77.

32b. Godson, G. N., Fiddes, J. C., Barrell, B. G., and Sanger, F., 1978, Comparative DNA sequence analysis of the G4 and X174 genomes, in: *The Single-Stranded DNA Phages*

(D. T. Denhardt, D. Dressler, and D. S. Ray, eds.), pp. 273–285, Cold Spring Harbor Laboratory, Cold Spring Harbor, New York.

33. Nowak, J. A., and Maniloff, J., 1979, Physical characterization of the superhelical DNA genome of an enveloped mycoplasmavirus, *J. Virol.* **29:**374–380.

34a. Matthews, R. E. F., 1982, Myoviridae, *Intervirology* **17:**68.

34b. Wood, W. B., and King, J., 1979, Genetic control of complex bacteriophage assembly, in: *Comprehensive Virology*, Vol. 13, (H. Fraenkel-Conrat and R. R. Wagner, eds.), p. 581, Plenum Press, New York.

34c. King, J., Hall, C., and Casjens, S., 1978, Control of the synthesis of page P22 scaffolding protein is coupled to capsid assembly, *Cell* **15:**551.

35. Kchromov, I. S., Sorotchkina, V. V., Nigmatullin, T. G., and Tikchonenko, T. I., 1980, A new nitrogen base 5-hydroxycytosine in phage N-17 DNA, *FEBS Lett.* **118:**51.

36. Calendar, R., Geisselsoder, J., Sunshine, M. G., Six, E. W., and Lindqvist, B. H., 1977, The P2-P4 transactivation system, in: *Comprehensive Virology*, Vol. 8 (H. Fraenkel-Conrat and R. R. Wagner, eds.), p. 329–344, Plenum Press, New York.

37. Eiserling, F. A., 1979, Bacteriophage structure, in: *Comprehensive Virology*, Vol. 13 (H. Fraenkel-Conrat and R. R. Wagner, eds.), p. 543, Plenum Press, New York.

38. Fuller, M. T., and King, J., 1981, Purification of the coat and scaffolding proteins from procapsids of bacteriophage P22, *Virology* **112:**529–547.

39a. Aposhian, H. V., 1975, Pseudovirions in animals, plants and bacteria, in: *Comprehensive Virology*, Vol. 5 (H. Fraenkel-Conrat and R. R. Wagner, eds.), p. 155, Plenum Press, New York.

39b. Thurm, Ph., and Garro, A. J., 1975, Bacteriophage-specific protein synthesis during induction of the defective *Bacillus subtilis* bacteriophage PBSX, *J. Virol.* **16:**179–183.

40. Matthews, R. E. F., 1982, Plasmaviridae, *Intervirology* **17:**55.

41. Matthews, R. E. F., 1982, Plectroviridae, *Intervirology* **17:**78.

42a. Schafer, R., and Franklin, R. M., 1975, Structure and synthesis of a lipid-containing bacteriophage. A. Polynucleotide-dependent polynucleotide-pyrophosphorylase activity bacteriophage PM2, *Eur. J. Biochem.* **58:**81–85.

42b. Mindich, L., 1978, Bacteriophages that contain lipid, in: *Comprehensive Virology*, Vol. 12 (H. Fraenkel-Conrat and R. R. Wagner, eds.), p. 271, Plenum Press, New York.

43. Matthews, R. E. F., 1982, Podoviridae, *Intervirology* **17:**70.

44. Coetzee, W. F., and Bekker, P. J., 1979, Plus-specific, lipid-containing bacteriophages PR4 and PR772: Comparison of physical characteristics of genomes, *J. Gen. Virol.* **45:**195–200.

45a. Spiegelman, S., Haruna, I., Holland, I. B., Beaudreau, G., and Mills, D., 1965, The synthesis of a self-propagating and infectious nucleic acid with a purified enzyme, *Proc. Natl. Acad. Sci. USA* **54:**919.

45b. Weissmann, C., Billeter, M. A., Goodman, H. M., Hindley, J., and Weber, H., 1973, Structure and function of phage RNA, *Annu. Rev. Biochem.* **42:**303–328.

46. Jeppesen, P. G. N., Argetsinger, S., Gesteland, R. F., and Spahr, P. F., 1970, Gene order in the bacteriophage R17 RNA: 5'-A protein-coat protein-synthetase-3', *Nature* **226:**230.

47. Khudyakov, I., Ya., Kirnos, M. D., Alexandrushkina, N. I., and Vanyushin, B. F., 1978, Yanophage S-2L contains DNA with 2,6-diaminopurine substituted for adenine, *Virology* **88:**8–18.

48. Rabussay, D., and Geiduschek, E. P., 1977, Regulation of gene action in the development of lytic bacteriophages, in: *Comprehensive Virology*, Vol. 8 (H. Fraenkel-Conrat and R. R. Wagner, eds.), pp 1–196, Plenum Press, New York.

49. Trautner, T. A., Pawlek, B., Günthert, U., Canosi, U., Jentsch, S., and Freund, M., 1980, Restriction and modification in *Bacillus subtilis*: Identification of a gene in the temperate phage SPβ coding for a *BsuR* specific modification methyltransferase, *Mol. Gen. Genet.* **180:**361–367.

50. Studier, F. W., 1972, Bacteriophage T7, *Science* **176:**367.

51. Matthews, R. E. F., 1982, Tectiviridae, *Intervirology* **17**:66.

52. Jones, P. T., and Pretorious, G. H. J., 1981, *Achromobacter* sp. 2 phage α3: A physical characterization, *J. Gen. Virol.* **53**:275–281.

53. Echols, H., and Murialdo, H., 1978, Genetic map of bacteriophage lambda, *Microbiol. Rev.* **42**:577.

54. Weisberg, R. A., Gottesman, S., and Gottesman, M. E., Bacteriophage λ: The lysogenic pathway, in: *Comprehensive Virology*, Vol. 8 (H. Fraenkel-Conrat and R. R. Wagner, eds.), pp. 197–258, Plenum Press, New York.

55. Howe, M. M., and Bade, E. G., Molecular biology of bacteriphage Mu. Genetic and biochemical analysis reveals many unusual characteristics of this novel bacteriophage, *Science* **190**:624.

56. Noyer-Weidner, M., Pawlek, B., Jentsch, S., Günthert, U., and Trautner, T. A., 1981, Restriction and modification in *Bacillus subtilis*: Gene coding for a BsuR-specific modification methyltransferase in the temperate bacteriophage φ3T, *J. Virol.* **38**:1077–1080.

57. Day, L. A., and Mindich, L., 1980, The molecular weight of bacterophage φ6 and its nucleocapsid, *Virology* **103**:376–385.

58a. Yoshikawa, H., Friedmann, T., and Ito, J., 1981, Nucleotide sequences at the termini of φ29 DNA (terminal-inverted repetition/linear DNA replication/early promoter sequences), *Proc. Natl. Acad. Sci. USA* **78**:1336.

58b. Anderson, D. L., Hickman, D. D., and Reilly, B. E., 1966, Structure of *Bacillus subtilis* bacteriophage φ29 and the length of φ29 deoxyribonucleic acid, *J. Bacteriol.* **91**:2081.

58c. Harding, N. E., Ito, J., and David, G. S., 1978, Identification of the protein firmly bound to the ends of bacteriophage φ29 DNA, *Virology* **84**:279–292.

59. Moynet, D. J., and DeFilippes, F. M., 1982, Characterization of bacteriophage φ42 DNA, *Virology* **117**:475–484.

60. Patel, J. J., 1976, Morphology and host range of virulent phages of lotus rhizobia, *Can. J. Microbiol.* **22**:204–212.

61. Einck, K. H., Pattee, P. A., Holt, J. G., Hagedorn, C., Miller, J. A., and Berryhill, D. L., 1973, Isolation and characterization of a bacteriophage of *Arthrobacter globiformis*, *J. Virol.* **12**:1031–1033.

62. Lake, J. A., 1974, Structure and protein distribution for the capsid of *Caulobacter crescentus* bacteriophage φCbK, *J. Mol. Biol.* **86**:499–518.

63. Denhardt, D. T., 1977, The isometric single-stranded DNA Phages, in: *Comprehensive Virology*, Vol. 7 (H. Fraenkel-Conrat and R. R. Wagner, eds.), pp. 1–104, Plenum Press, New York.

64. Ackermann, H. W., *et al.*, 1976, Structural Aberrations in Group A Staphylococcus Bacteriophages, *J. Virol.* **18**:619–626.

65. Moazamie, N., Ackermann, H.-W., and Murphy, M. R. V., 1979, Characterisation of two Salmonella Newport bacteriophages. *Can. J. Microbiol.* **25**:1063–1072.

66. Schechtman, M. G., Snedeker, J. D., and Roberts, J. W., 1980, Genetics and structure of the late gene regulatory region of phage 82, *Virology* **105**:393–404.

67. Pretorius, G. H. J., and Coetzee, W. F., 1980, *Proteus mirabilis* phages 5006M, 5006M HFT *k* and 5006 M HFT *ak*: Physical comparison of genome characteristics, *J. Gen. Virol.* **49**:33–39.

Figures

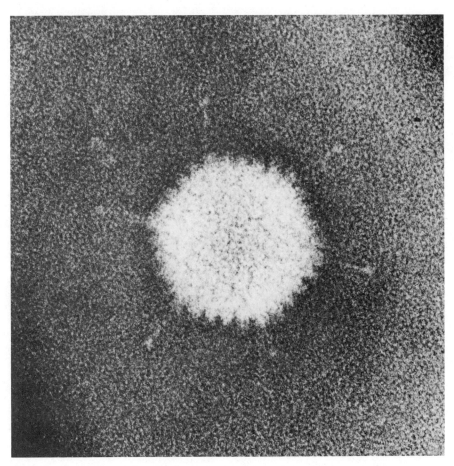

FIGURE 1. Adenovirus (80 nm diameter largely composed of hexons, with the fibers attached to the pentons being 27-nm long).

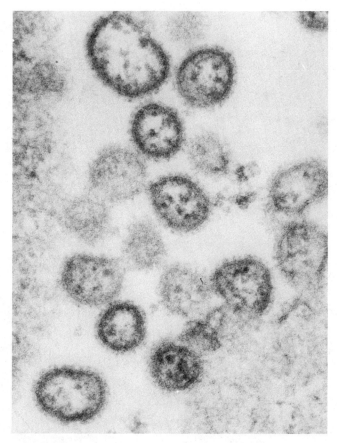

FIGURE 2. Lymphocytic choriomeningitisvirus (Arenaviridae). Typical diameter 150 nm. The small internal bodies are ribosomes.

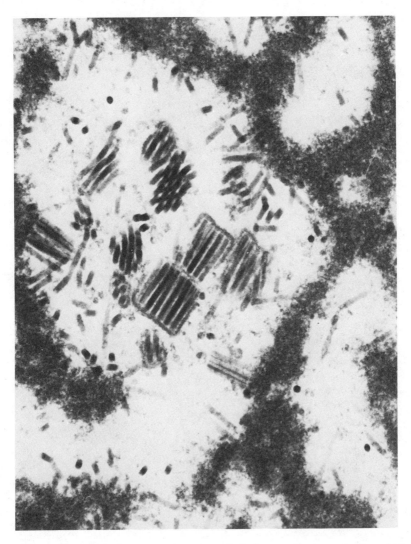

FIGURE 3. Nuclear polyhedrosis virus (Baculoviridae). Groups of virion rods (about 300-nm long) more or less orderly stacked and encapsidated.

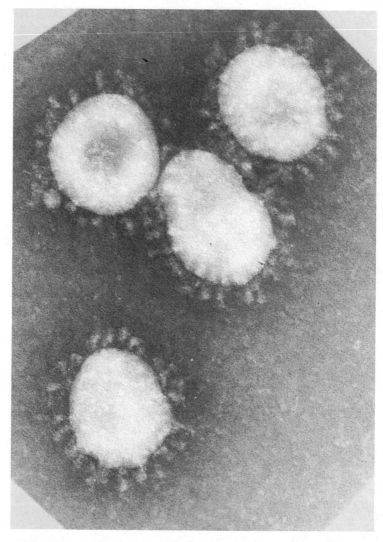

FIGURE 4. Human coronavirus (Coronaviridae). Average diameter 150 nm. The long peplomers are clearly visible.

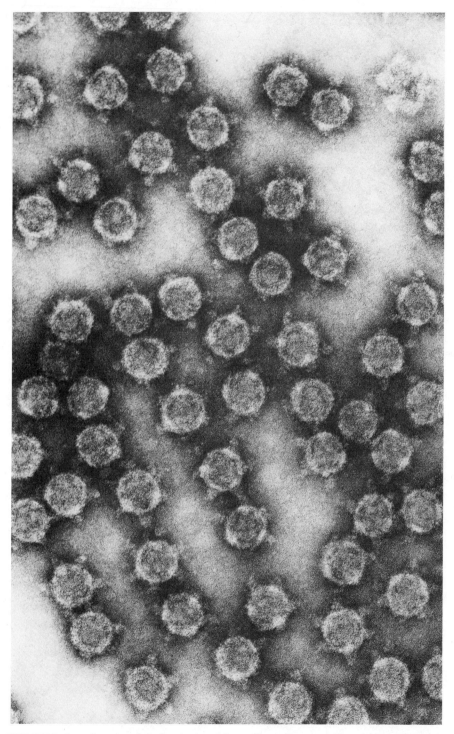

FIGURE 5. Cytoplasmic polyhedrosis virus (about 60 nm diameter), (Reoviridae). The tube-like protrusions through which the RNA molecules exit upon translation and replication are clearly visible.

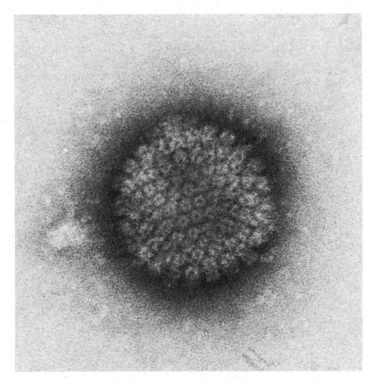

FIGURE 6. Human herpesvirus (about 200 nm diameter). The tubular capsomers, 162 in all, are clearly distinguishable.

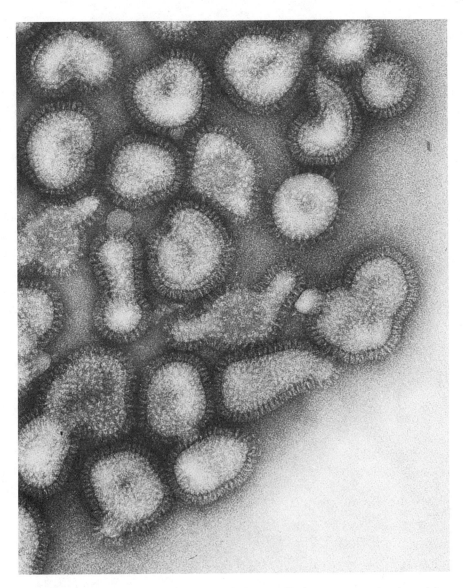

FIGURE 7. Influenzavirus (Orthomyxoviridae), typically heterogeneous in shape (pleomorphic), average diameter 120 nm.

FIGURE 8. Tipula iridescent virus (top) (180 nm diameter), (Iridoviridae) shadowed by the electron beam, with an icosahedral model body (bottom) equally light-shadowed to illustrate the three-dimensional structure of a large icosahedral viruses. Most small isometric viruses have the same shape, not usually so clearly visible.

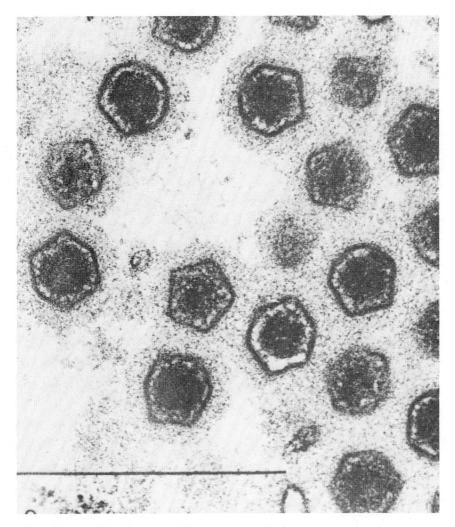

FIGURE 9. Lymphocystis virus (fish virus, Iridoviridae), (250 nm diameter); group formerly termed icosahedral cytoplasmic DNA viruses.

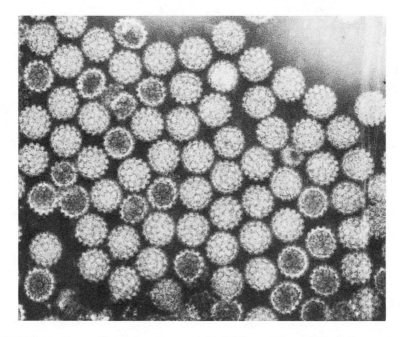

FIGURE 10. Human papillomavirus (Papovaviridae), (55 nm diameter). The polyoma viruses, e.g., SV40, look very similar, though slightly smaller (45 nm diameter).

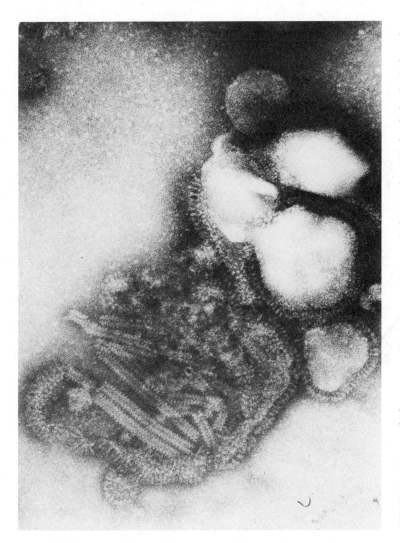

FIGURE 11. Newcastle disease cirus (Paramyxoviridae), partly fractured to visualize helical rod-shaped nucleocapsid fragments.

FIGURE 12. Poliovirus. A microscopic crystal of the pure virus (top), and an electron micrograph of the crystalline array of the virions (bottom) (27 nm diameter).

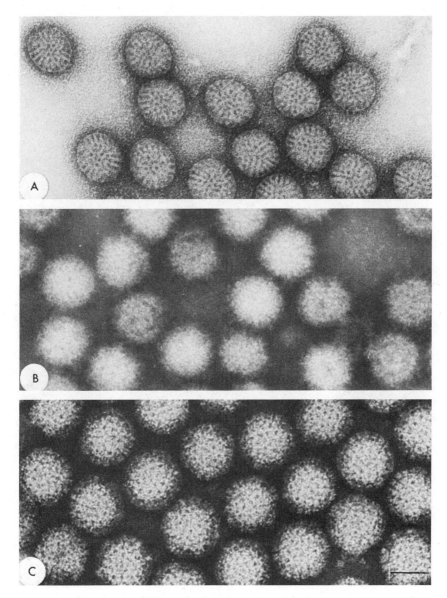

FIGURE 13. Three genera of Reoviridae (see also Figure 5), orthoreovirus (C), orbivirus (B), and rotavirus (A). The tubelike capsomers are most clearly seen in C. Diameter about 80 nm.

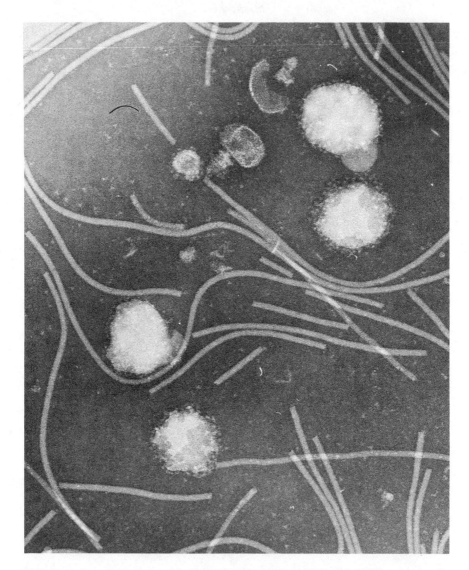

FIGURE 14. Rous sarcoma virus (average diameter 100 nm). The thread-like particles represent potatovirus X used as size markers (13 nm diameter and usually about 500-nm long).

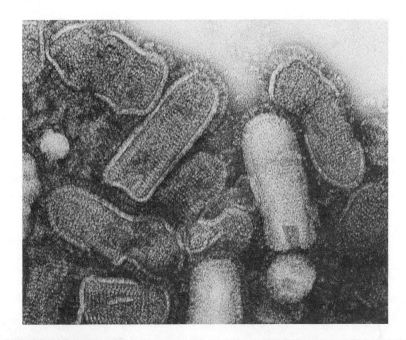

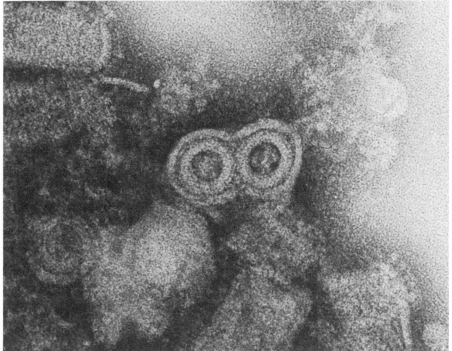

FIGURE 15. Vesicular stomatitis virus (Rhabdoviridae) as typically seen bullet-shaped particles (60 × 200 nm) (top), and from the end which shows the glycoprotein envelope and the matrix coat (bottom).

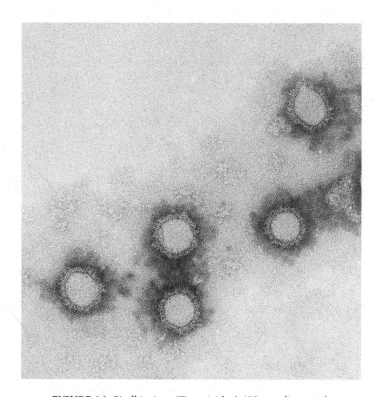

FIGURE 16. Sindbisvirus (Togaviridae), (30 nm diameter).

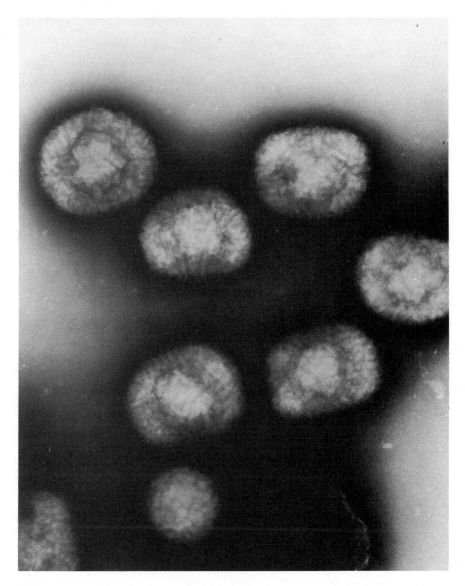

FIGURE 17. Vacciniavirus (Poxviridae), (about 400 × 240 × 200 nm). The internal lateral bodies and other structural elements are discernable.

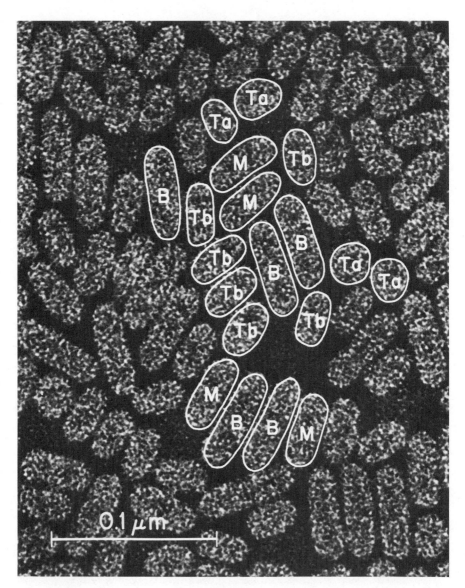

FIGURE 18. Alfalfa mosaic virus. Four different particles of this tripartite virus are illustrated as follows: B (bottom upon ultracentrifugation) the largest, M (middle), Tb (top), all nucleoproteins, and Ta, RNA-lacking pseudovirions. (58, 48, 36, and 28 × 28 nm).

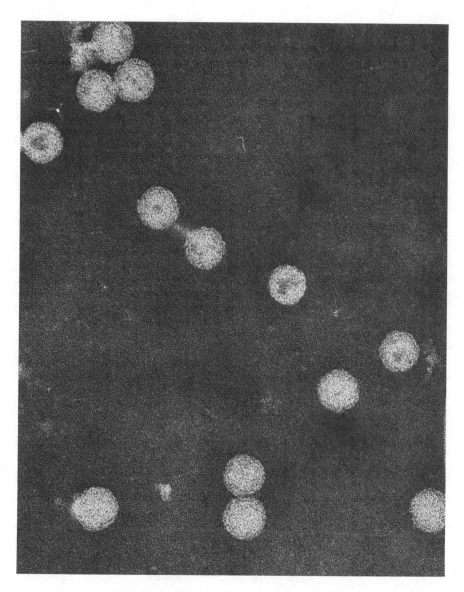

FIGURE 19. Cauliflower mosaic virus (50 nm diameter).

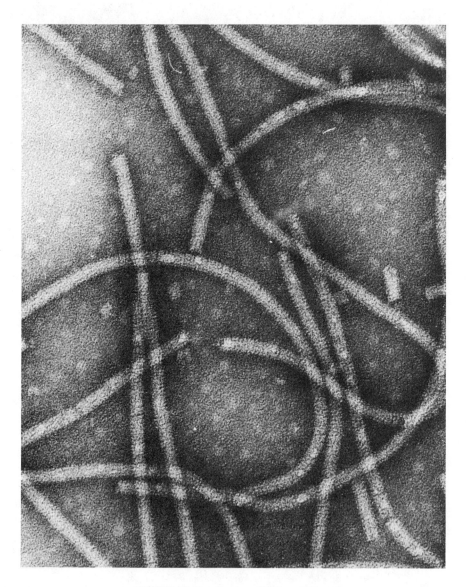

FIGURE 20. Potato virus X (13 × 600 nm).

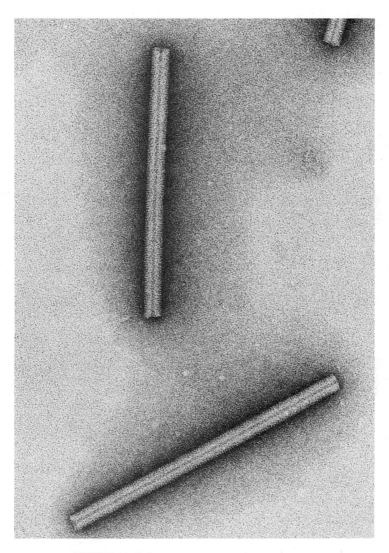

FIGURE 21. Tobacco mosaic virus (18 × 300 nm).

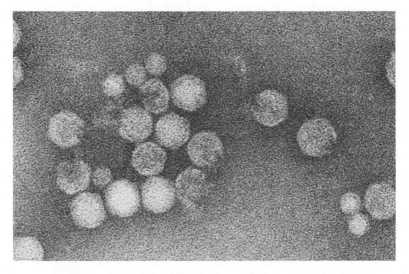

FIGURE 22. Tobacco necrosis virus and its satellite virus (28 and 17 nm diameter, respectively).

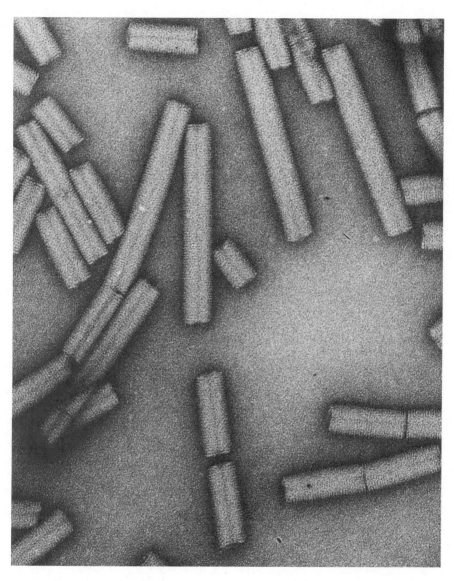

FIGURE 23. Tobacco rattle virus. Typical particles of this strain are 200- and 80-nm long.

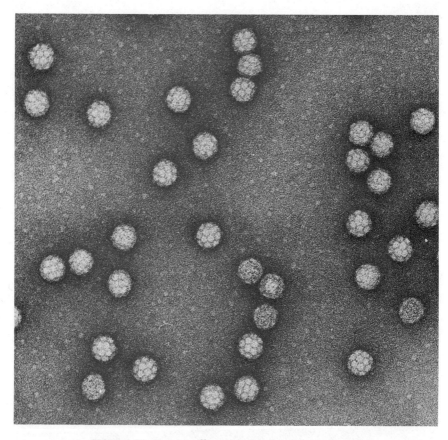

FIGURE 24. Turnip yellow mosaic virus (29 nm diameter).

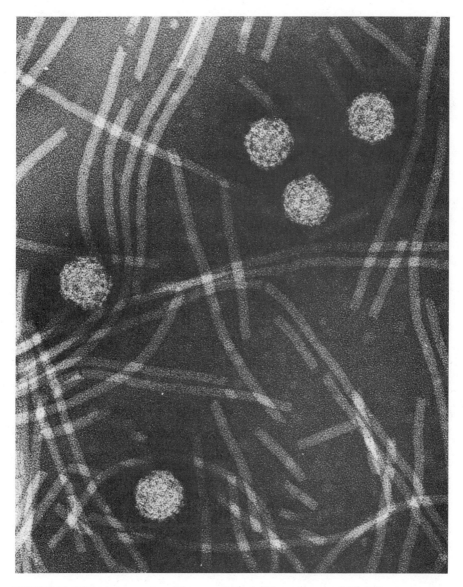

FIGURE 25. Wound tumor virus (Phytoreoviridae), (75 nm diameter). The fibers are potato virus X.

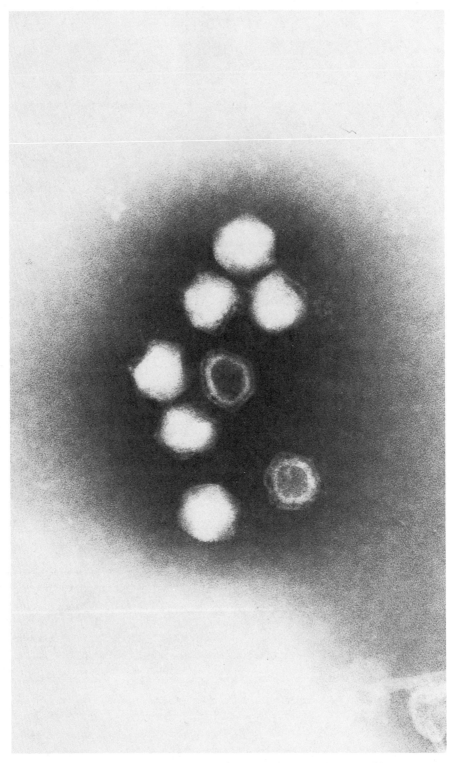

FIGURE 26. Phage B am 35 (63 nm head, with spikes at vertices and short tail) (Tectiviridae).

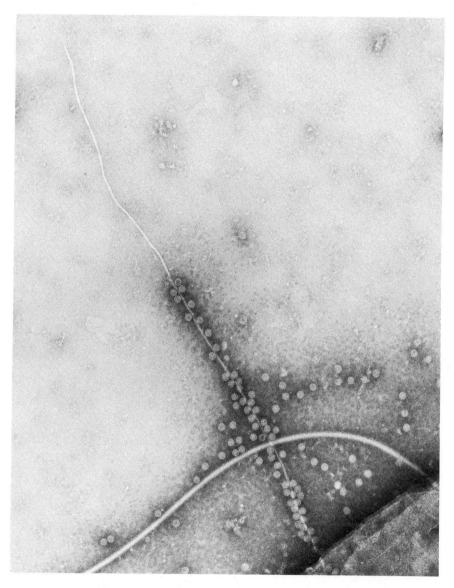

FIGURE 27. Phage fd (Inoviridae, group I), attached to the end of a pilus of *Escherichia coli*, with many MS2 phages (Leviviridae, 23 nm diameter) attached to the side of that pilus.

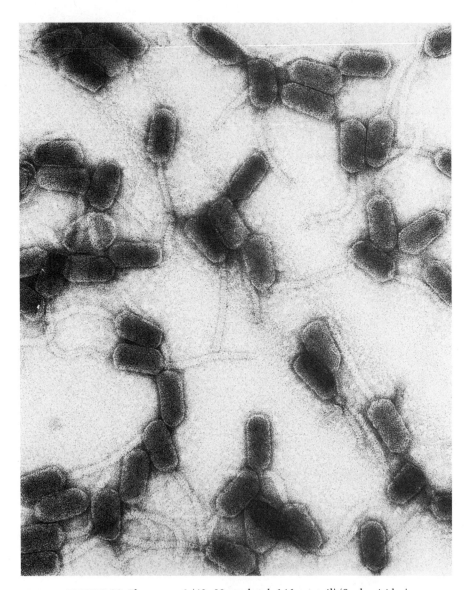

FIGURE 28. Phage *mor* 1 (43–88 nm head, 146 nm tail) (Syphoviridae).

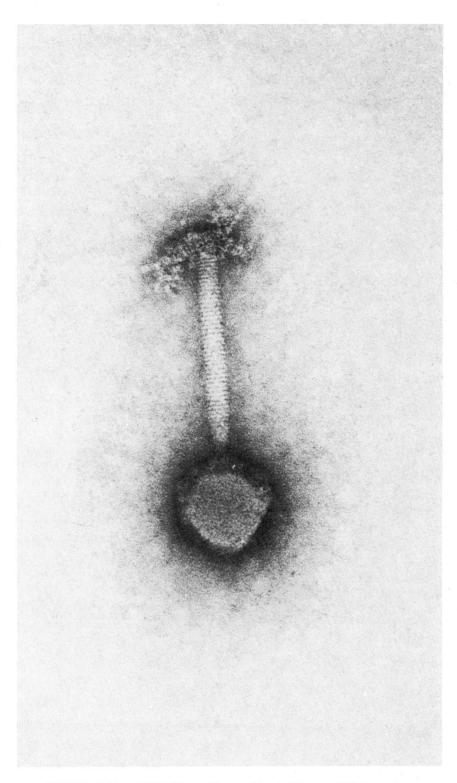

FIGURE 29. Phage MP13 (Note odd-shaped head; 200 nm tail) (Myoviridae).

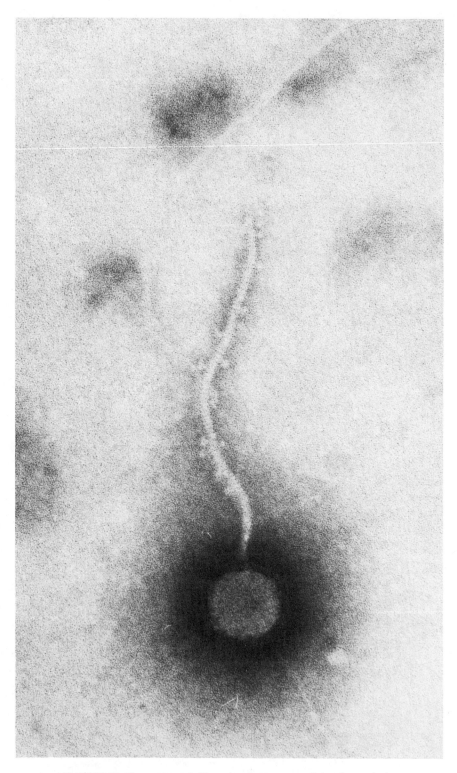

FIGURE 30. Phage MP15 (Oblong head, 300 nm tail) (Syphoviridae).

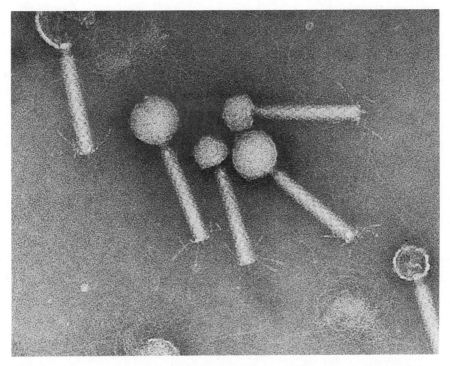

FIGURE 31. Phage P2 (60 nm diameter) and dependent P4 phage with smaller head (45 nm diameter). The tails of both are 135 nm long (Myoviridae).

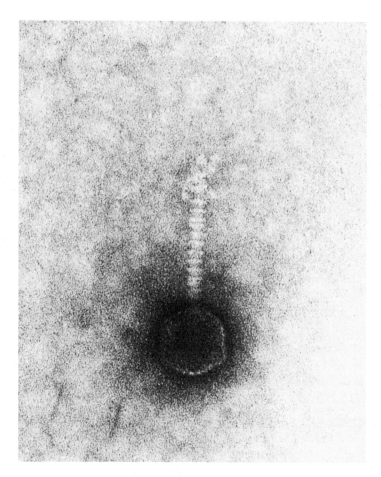

FIGURE 32. Phage Rhiφ I (60 nm head, 130 nm tail) (Myoviridae).

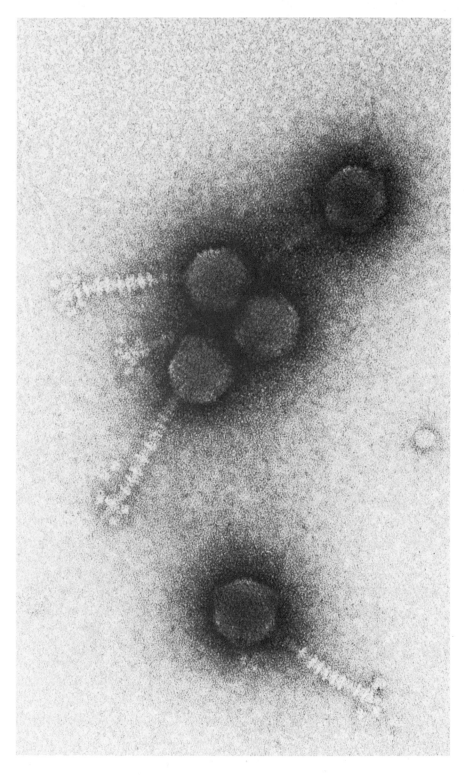

FIGURE 33. Phage Rhiφ L9 (50 nm head, 120 nm tail; note complex tail appendages) (Myoviridae).

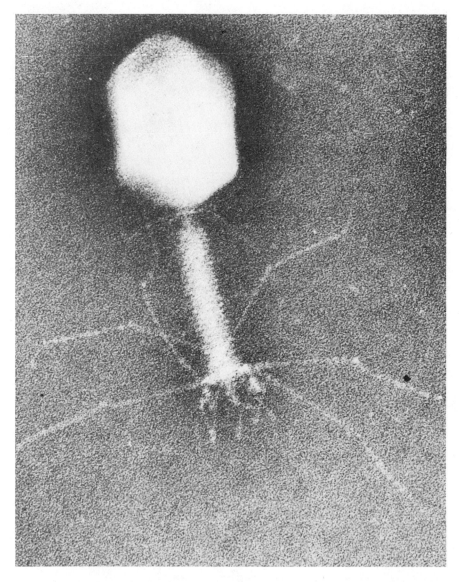

FIGURE 34. T4 phage (80–95 nm head and 16 × 110 nm tail), with clearly visible neck, tail fibers, base plate, and spikes (Myoviridae).

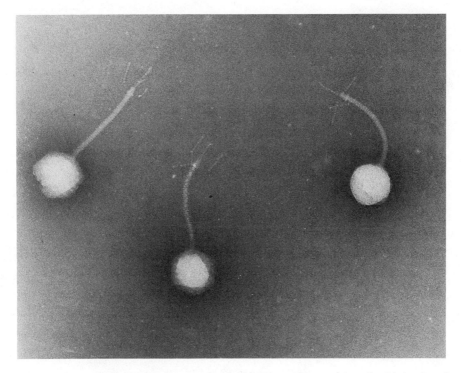

FIGURE 35. T5 phage (65 nm diameter head and 110 × 180 nm thin tail with knob and fibers at the end (Styloviridae).

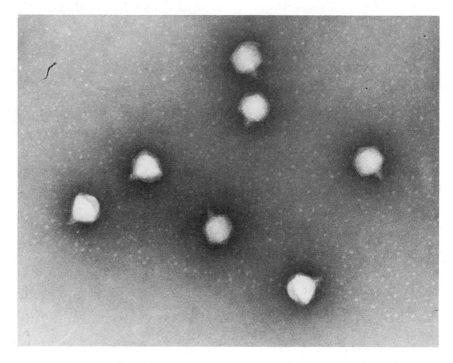

FIGURE 36. T7 phage (65 nm diameter with very thin 17 nm tail) (Podoviridae).

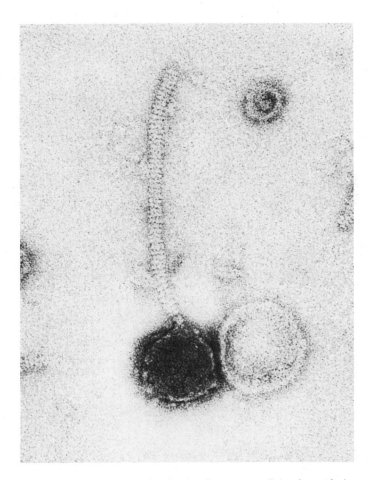

FIGURE 37. Phage VP 6 (69 nm head, 233 nm tail) (Syphoviridae).

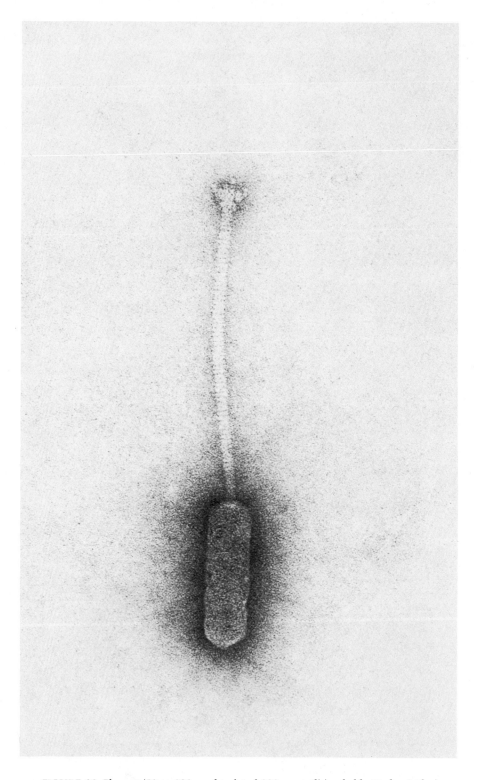

FIGURE 38. Phage θ (50 × 100 nm head and 200 nm tail) (probably Syphoviridae).

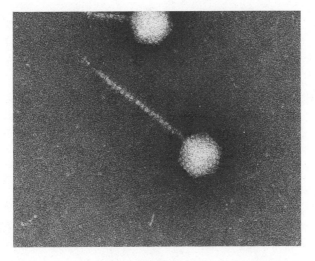

FIGURE 39. Phage λ (54 nm diameter head, 15 × 150 nm tail) (Syphoviridae).

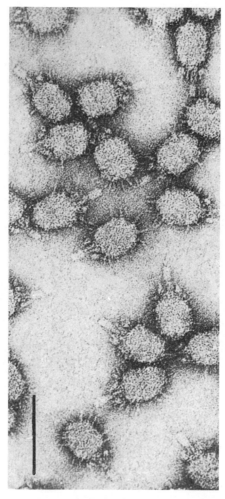

FIGURE 40. Phage φ 29 (32 × 46 nm head, 6 × 36 nm tail, with fibers on head and collar).

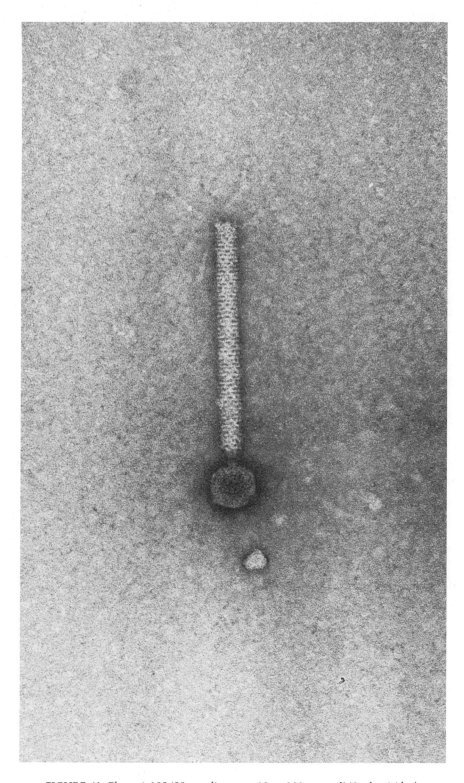

FIGURE 41. Phage φ 105 (59 nm diameter, 10 × 200 nm tail) (Syphoviridae).

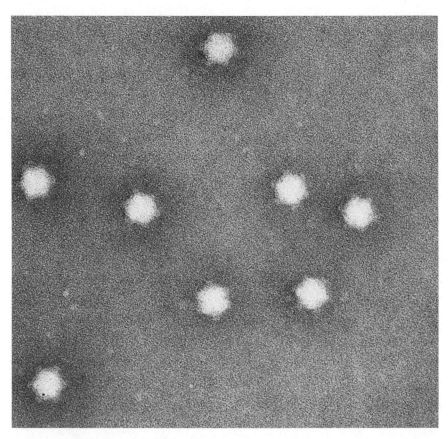

FIGURE 42. Phage φ χ 174 (27 nm diameter, with spikes at the vertices of the icosahedron) (Mycroviridae).

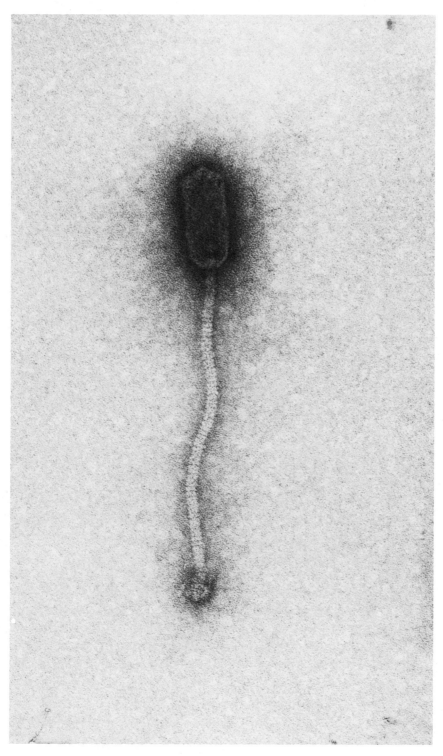

FIGURE 43. Phage 6 (40 × 92 nm head and 300 nm tail) (Syphoviridae).

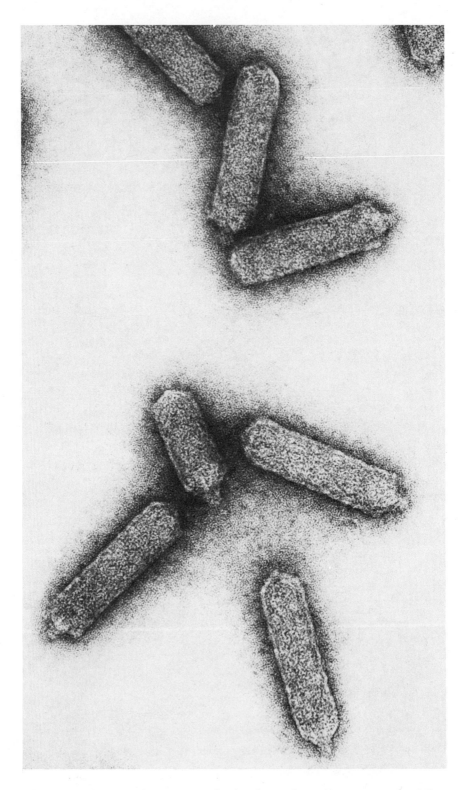

FIGURE 44. Phage 7-11 (40 × 154 nm head and very short tail; note presence of shorter particles) (Podoviridae).